SpringerBriefs in Applied Sciences and Technology

SpringerBriefs present concise summaries of cutting-edge research and practical applications across a wide spectrum of fields. Featuring compact volumes of 50 to 125 pages, the series covers a range of content from professional to academic.

Typical publications can be:

- A timely report of state-of-the art methods
- An introduction to or a manual for the application of mathematical or computer techniques
- A bridge between new research results, as published in journal articles
- A snapshot of a hot or emerging topic
- An in-depth case study
- A presentation of core concepts that students must understand in order to make independent contributions

SpringerBriefs are characterized by fast, global electronic dissemination, standard publishing contracts, standardized manuscript preparation and formatting guidelines, and expedited production schedules.

On the one hand, **SpringerBriefs in Applied Sciences and Technology** are devoted to the publication of fundamentals and applications within the different classical engineering disciplines as well as in interdisciplinary fields that recently emerged between these areas. On the other hand, as the boundary separating fundamental research and applied technology is more and more dissolving, this series is particularly open to trans-disciplinary topics between fundamental science and engineering.

Indexed by EI-Compendex, SCOPUS and Springerlink.

Norashikin M. Thamrin · Addie Irawan ·
Zurita Zulkifli · Syed Abdul Mutalib Al Junid ·
Megat Syahirul Amin Megat Ali ·
Anwar P. P. Abdul Majeed

Robotics in Education

Springer

Norashikin M. Thamrin
Microwave Research Institute
Universiti Teknologi MARA
Shah Alam, Selangor, Malaysia

Zurita Zulkifli
Faculty of Electrical Engineering
Universiti Teknologi MARA
Shah Alam, Selangor, Malaysia

Megat Syahirul Amin Megat Ali
Microwave Research Institute
Universiti Teknologi MARA
Shah Alam, Selangor, Malaysia

Addie Irawan
Faculty of Electrical and Electronics
Engineering Technology (FTKEE)
Universiti Malaysia Pahang Al-Sultan
Abdullah
Pekan, Pahang, Malaysia

Syed Abdul Mutalib Al Junid
Faculty of Electrical Engineering
Universiti Teknologi MARA
Shah Alam, Selangor, Malaysia

Anwar P. P. Abdul Majeed
Faculty of Engineering and Technology
Sunway University
Subang Jaya, Malaysia

ISSN 2191-530X ISSN 2191-5318 (electronic)
SpringerBriefs in Applied Sciences and Technology
ISBN 978-981-95-3547-7 ISBN 978-981-95-3548-4 (eBook)
https://doi.org/10.1007/978-981-95-3548-4

This Springer imprint is published by the registered company Springer Nature Singapore Pte Ltd.
The registered company address is: 152 Beach Road, #21-01/04 Gateway East, Singapore 189721, Singapore

Preface

In a world increasingly shaped by automation, artificial intelligence, and connected systems, robotics has emerged as a vital component in education, one that transcends the traditional boundaries of STEM disciplines. This book, *Robotics in Education*, is a curated collection of research, practices, and reflections from diverse contributors who each bring unique perspectives on how robotics can empower both teaching and learning. Rather than presenting a unified narrative, the chapters are intentionally varied, showcasing the multidimensional nature of robotics education across different contexts, age groups, and levels of technical sophistication.

The importance of research in robotics education cannot be overstated. As classrooms evolve into more interactive, interdisciplinary environments, educators and learners alike are seeking tools that foster engagement, autonomy, and critical thinking. Robotics offers all of these and more. It enables experiential learning where students can design, build, and iterate systems that mirror real-world challenges. Whether applied at the primary level through simple programmable platforms or in higher education using AI-driven systems, robotics creates opportunities for learners to merge theory with tangible action.

This book addresses the growing global demand for guidance on integrating robotics into educational settings. The escalation in robotics education research is fuelled not only by technological advancements but also by a more profound pedagogical shift, the one that values creativity, problem-solving, and collaboration as essential skills for the future. With robotics as a medium, education becomes a launchpad for innovation rather than a container of static knowledge.

What makes this volume unique is its breadth and accessibility. The contributors do not simply present theoretical discourse; they offer hands-on strategies, real case studies, and grounded reflections that can be adapted to diverse learning environments. The book is designed for educators, researchers, curriculum designers, and anyone interested in reimagining what modern education can look like when infused with intelligent systems.

By reading this book, readers will gain practical insights, research-based frameworks, and inspiration to implement robotics in ways that are pedagogically sound, technically feasible, and deeply impactful for learners. It invites readers not only to understand robotics education but also to shape it.

Shah Alam, Malaysia — Norashikin M. Thamrin
Pekan, Malaysia — Addie Irawan
Shah Alam, Malaysia — Zurita Zulkifli
Shah Alam, Malaysia — Syed Abdul Mutalib Al Junid
Shah Alam, Malaysia — Megat Syahirul Amin Megat Ali
Subang Jaya, Malaysia — Anwar P. P. Abdul Majeed

Contents

Robotics in Education—Transforming Stem Learning and Skill Development

1 Introduction

The heightened utilisation of robots across diverse industrial sectors, including agricultural, health, automotive, and manufacturing, has advanced the discipline of robotics significantly, making it a crucial component of educational curricula [1–5]. The industrial revolution (IR4.0) emphasises the integration of Science, Technology, Engineering, and Mathematics (STEM) into the educational sector at both primary and tertiary levels. The formulation of a curriculum that introduces innovations in the education system is essential to guarantee that students receive STEM-based information. The study of robotics is now regarded as an essential element of education rather than a specialised hobby.

1.1 Overview of Robotics in Education

The implementation of robotics in education has established a connection between theory and practice. The disciplines of mathematics, physics, computer science, and engineering can be articulated through robotic applications. Valuing knowledge is enhanced by the interplay between theory and practice. This program can cultivate students who are critical and creative in problem-solving, logical reasoning, and design skills, which are essential for developing talent with a STEM education background.

Robotics is no longer perceived as a challenging discipline, as numerous platforms, like Micro:bit, Arduino, and LEGO Mindstorm, facilitate students in robot creation. These existing technologies facilitate the learning of robotics and require only a minimal background. Generally, individuals can acquire knowledge in robotics with basic necessities such as a computer, internet access, and electronic components. At the tertiary education level, this expertise in robotics has been integrated

N. M. Thamrin et al., *Robotics in Education*,
SpringerBriefs in Applied Sciences and Technology,
https://doi.org/10.1007/978-981-95-3548-4_1

with artificial intelligence, complicated simulations, and machine learning for more sophisticated robotic applications [6–8].

The government's role in emphasising robotics instruction within the school curriculum is perceived as a means of fostering and enhancing computer skills and digital literacy among students. Since 2017, Malaysia has implemented the topic of design technology (RBT) and the fundamentals of computer science (ASK) in schools to promote STEM education.

The participation of the government, non-governmental organisations (NGOs), and business entities in organising robotics competitions has fostered student enthusiasm and enhanced robotics expertise in education. Certain schools have established robotics clubs to offer a more concentrated environment for interested students and to enhance the robotics ecosystem through collaboration across schools, universities, government entities, NGOs, and business organisations.

The advancement of robotics knowledge is not restricted to the affluent; the accessibility of open-source platforms online enables underserved communities to attain equity in robotics education. As an example, the triumph of indigenous children in a robotics competition has enabled underserved youth to achieve success on an international scale in Busan, Korea [9].

1.2 Importance of Robotics in the Twenty-First Century Educational Landscape

Robotics education is an important field that improves the parts of Science, Technology, Engineering, and Mathematics (STEM). Learning in the twenty-first century focuses on skill-based education that changes quickly to keep up with new technologies. This type of learning encourages critical thinking, problem-solving, teamwork, and adaptability. Robotics education meets all the requirements for twenty-first-century learning [10–13]. For example, students need to be able to think creatively and have the skills to solve problems. Robotics education gives students the abilities they need to face difficulties in the future, setting them up for success in the modern world.

Some ways to include robotics lessons in the education landscape are:

(1) Improving STEM through hands-on and interactive learning activities

Robotics turns abstract STEM ideas into real-life experiences. Instead of just learning theory, students design, build, and program robots, which lets them see how physics, math, and engineering principles can be used in real life. This hands-on approach keeps students interested and helps them understand concepts better, especially those who have trouble with traditional methods [14].

(2) Improving Higher-Order Thinking Skills (HOTS)

Students often need to use critical thinking, creativity, curiosity, and logical reasoning to solve robotic problems. Students learn how to solve problems well while they debug and improve their designs. These skills are very important for dealing with complicated situations in today's jobs.

(3) Encouraging working together and communicating with each other

Most of the time, robotics projects require students to work together, share ideas clearly, and delegate tasks effectively. These experiences help people learn how to work together, be a good leader, and communicate clearly, all of which are important abilities for the twenty-first-century workplace.

(4) Building skills in programming and digital literacy

Programming is a basic skill for working with robots. Giving students coding abilities gives them important digital literacy and programming skills that are particularly useful in fields like robotics, AI, cybersecurity, and data science [15–17].

(5) Supporting new ideas and business-mindedness

Robotics encourages people to be curious and creative. Open-ended challenges encourage students to come up with new ideas and build prototypes of their innovations. This helps students develop an entrepreneurial mindset, which prepares them to not only get jobs but also become the next generation of tech innovators and problem-solvers [18].

(6) Making STEM more accessible and diverse

Robotics is becoming a powerful tool for making STEM education more accessible to everyone. This is thanks to the growing availability of affordable kits and online materials that can engage students with different backgrounds and learning styles. It helps close the gap in access to education between rural and urban areas, especially when done through outreach and community programs.

2 Integration of STEM in Curriculum

The four pillars of science, technology, engineering, and mathematics are the foundation of robotics. By cultivating students' collaborative abilities, students can effortlessly acquire a comprehensive understanding of robotics by mastering the fundamentals of this STEM subject. The formulation of strategies through various models will facilitate the attainment of twenty-first-century learning, which encompasses problem-based solving, critical thinking, cooperation, collaboration, creative and innovative ideation. Research has demonstrated that the integration of robotics into the educational process can have a positive impact on the academic and social abilities of students. Several strategies have been implemented to evaluate the efficacy of incorporating robotics into STEM curriculum learning.

2.1 Methods for Integrating Robotics into the STEM Curriculum:

(1) Professional development of educators in the field of STEM pedagogy

It is crucial to implement strategies that enhance STEM pedagogy among instructors in order to guarantee that they possess the necessary expertise to deliver STEM in the classroom. One of the activities that can be implemented to enhance STEM pedagogy is the periodic conduct of STEM-focused seminars that address the areas of inquiry-based learning, engineering design, coding, and the integration of real-world problem-solving. Furthermore, the implementation of mentorship programs to assist new teachers and those without a STEM background. By monitoring, evaluating, and providing feedback on STEM learning, educators who possess a robust STEM pedagogy foundation can facilitate and ease the progression of the STEM curriculum [19]. Additionally, educators may engage in partnerships with universities, industry, and professional organisations to integrate robotics into the curriculum. Developing micro-credentials and obtaining recognised certificates in STEM are opportunities that teachers can take advantage of to enhance their continuous motivation.

(2) Have the opportunity to work together with STEM laboratories or Maker Spaces

The STEM learning curriculum can be fulfilled through the planning of a variety of activities in STEM laboratories and Maker Spaces. Excellent collaboration can result in cost savings, particularly when it comes to robots and their components. Teachers have the opportunity to collaborate in the development of STEM modules in the context of module preparation. STEM Labs that can be established in schools will be capable of educating students on the organisation of programs, the management of STEM devices inventory, including microcontrollers and discrete components, and enhancing the effectiveness of schedule management [20].

(3) Evaluation through design thinking and problem-solving skills

When robotics is implemented in STEM education, educators have the option of turning from conventional assessments to problem-solving and design thinking strategies. Students can acquire twenty-first-century skills through teamwork, critical and creative thinking.

(4) Robotics in Science, Technology, Engineering, Arts, and Mathematics (STEAM): Cross-disciplinary approaches

The study of robotics is not defined by the integration of Science, Technology, Engineering, and Mathematics. In reality, the demands of the art subject are highly beneficial to the development of a robot [21–23]. Students acquire a more comprehensive understanding of the subject matter when the curriculum is designed to incorporate a variety of disciplines. Robotics projects that are themed around soccer robots, robots that assist patients, robots that are used in agriculture and other topics can be developed by educators. Students can incorporate knowledge from a variety of perspectives by exploring these themes. For instance, they can use mathematics to

Fig. 1 Shows a group of students measuring the distance to estimate the position of the drone landing location

estimate the size of robots, art subjects to design robots, and computer basics to program robots according to their functions. In order to ensure that all robot designs are evaluated equally in this curriculum, educators may establish a rubric. Figure 1 shows participants using a measuring tape to measure the distance to ensure that the robot could reach its destination accurately.

(5) Promoting involvement in robotics competitions and exhibitions

Students will be motivated to exhibit a high level of competitiveness and enthusiasm by the school's support of the robotics curriculum through participation in competitions organised by the Education Department or external parties. Participation in robotics competitions can also enhance students' social skills, in addition to cultivating critical thinkers. Participation in robotics will facilitate students' comprehension of STEM concepts. Figure 2 shows an autonomous line tracer robot.

Fig. 2 Line tracer robot

2.2 *Curriculum Design: The Role of Robotics as a Core or Supplementary Component*

In the development of a robotics curriculum, whether it is a core or supplementary component, the fundamental requirements for robotics learning, which are robotics-related skills, must be attained as a course outcome of the robotics subject. In addition to acquiring literacy and numeracy skills in other courses, students must also master problem-solving, design thinking, and coding in robotics. The 6E model (engage, investigate, explain, engineer, enrich, and evaluate steps) is the foundation of an effective curriculum that aims to develop robotics-related skills [24].

2.2.1 Robotics as a Core Subject

One of the core subjects that can be incorporated into the school curriculum is robotics. For instance, the Montessori Early Childhood Classroom has implemented the robotics module as an integral component of its curriculum. Collaboration with robotics platforms such as Arduino, LEGO, and others can facilitate the adaptation of the robotics module from established modules at the university or maker lab. The organisation of learning should be based on sub-modules that can be implemented during the academic year [25]. Alam and Mohanty introduced Integrated Constructive Robotics in Education (ICRE) to demonstrate that robotics education has prioritised student-centred, hands-on experiences, which enable students to experimentally implement, explore, and apply concepts in real-time [26]. The advantage of embedding robotics as a core subject in the school curriculum is that everyone is encouraged towards robotic education. The curriculum becomes more pertinent to the technological landscape through robotic education, which incorporates contemporary technologies such as artificial intelligence (AI), augmented reality (AR), and virtual reality (VR).

2.2.2 Robotic as a Supplementary Course

In addition to the core curriculum, robotics may be provided as a supplementary course, with participation either optional or limited to students who show interest. Collaboration and engagement with the community are essential. Practical relationships, mentorship, and insights into robotics applications in professional environments are accessible through collaborations with industry, educational institutions, and the community. Fostering an environment that encourages student engagement outside of the classroom is facilitated by engaging parents and other communities in robotics initiatives. The execution of the informal robotic program is often conducted during school holidays, offered as an elective, after-school clubs, summer camps, or self-learning at a maker hub or any private STEM-related organisation [27]. Unlike robotics education, which is a part of the school curriculum, the program has a shorter duration. For example, the program may be offered weekly on weekends, for 4–6 weeks during school holidays, or as a daily program organised by the robotics program coordinators. The flexible schedule and implementation of this elective robotic education are among its advantages. Furthermore, programs such as this enable individuals who are interested in learning more in-depth to do so without being weighed down by evaluation and assessment, as in the core curriculum. Nevertheless, students are required to pay in order to participate in certain informal robotic programs. The obstacle is that not all individuals are able to participate in the robotic programs unless they receive financial and equipment support from individuals, industries, universities, governments, or NGOs.

3 Skill Building Through Robotics

Automation technology is advancing swiftly in the contemporary environment. To enhance understanding in the realm of automation, students must acquire skills such as technical proficiency (digital literacy), critical thinking, problem-solving, and creativity to remain competitive in the rapidly modernising world. These future talents can be implemented through robotics education, as seen in Fig. 3. The enhancement of these skills enables students to acquire twenty-first-century skills.

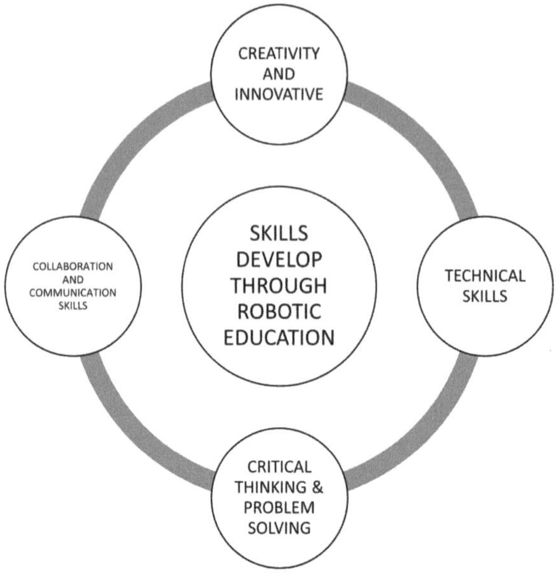

Fig. 3 Twenty-first century skills adapted from robotic education

3.1 Development of Technical Skills: Programming, Engineering, and Design

Robotics education cultivates technical skills, including ability in coding. Diverse software applications might be utilised to educate individuals in programming proficiency. Numerous open-source software options are available for free, including MakeCode, Scratch, and Mblock, suitable for elementary school kids, while programming languages such as C, C++, and Python are appropriate for high school and university students. Programming is essential for providing instructions to robots based on the necessary functions. This talent, referred to as digital literacy, encompasses proficiency in utilising computers, comprehending syntax, and grasping programming languages.

In addition to programming proficiency, comprehending engineering principles, including various microcontroller types and their architectures, as well as the functions and characteristics of sensors, is essential for robotics education. Students can develop several types of robotic functions, each with unique designs, utilising 3D design software such as AutoCAD, TinkerCad, and others. Skill in software for robotic design can enhance students' creativity and ingenuity.

3.2 Enhancing Critical Thinking and Problem-Solving Abilities

Prior to initiating a project, the problem statement is defined via the design thinking process to attain the specified objectives. Likewise, prior to initiating the design process of a robot, an in-depth analysis of the problem statement must be conducted. To cultivate critical thinking, students must engage in analysis, evaluation, and creation, rather than only remembering information. This structured and methodical process will yield a more organised and systematic product. Critical thinking instructs individuals to consistently inquire "why," "what if," and to embrace the possibility of failure. The opportunity for self-improvement and continual effort is perpetually available. Project-based learning is an exemplary way for assessing open-ended problem-solving abilities [28, 29]. There is no singular correct answer; instead, it invites students to engage in critical thinking to devise solutions to the problem statement. This strategy cultivates a student's independence, confidence, and courage to experiment. Educators can assist students in improving their critical thinking by employing instructional tools such as deconstructing the problem statement, creating mind maps, and utilising cause-and-effect charts that explain the details of the problem to be addressed.

3.3 Encouraging Creativity and Innovation with Robotics Competitions

A method to foster student creativity is to present difficulties as open-ended tasks. A step-by-step instruction for students to resolve them is omitted. It must originate from student brainstorming to address the specified issue. As an example, develop a robot that capable to assist immobile persons in taking medications. Students need to think about the design of the robot, the need for the robot to have a place to put medicine, have a shifter to go up and down so that bedridden patients can take medicine according to the height of their bed and so on. The contextual aspects must be listed prior to the creative development of the robot design. All aspects must be thoroughly investigated to enhance clarity, creativity, and innovation in robot design production.

3.4 Collaboration and Communication Skills in Robotics Projects

Teamwork is essential when addressing problems or developing robots. Effective communication is a key component of successful teamwork. During brainstorming, students must generate ideas, resolve problems, convey technical concepts, and learn how to cooperate with parties other than group members. Robotics education is much more than just providing students with technical skills; it also includes developing students' soft skills. Robotics development also prepares students to be responsible for assigned responsibilities such as programmer, designer, documenter, project manager, and presenter. If one of the tasks is not done successfully, the entire project will fail. That is why communication and collaboration are so crucial, and they may be developed through robotics education.

3.5 Challenges and Obstacles in Introducing Robotic Education

Although robotic education has the potential to teach twenty-first-century abilities, there are a number of hurdles and obstacles that continue to impede its development in schools. Among these include finances, a lack of qualified teachers, a lack of room to build robotic laboratories, a lack of support from school administration and parent organisations, and a lack of student engagement. Teacher assistance is critical in ensuring that children have access to robotics education.

4 Conclusion

Robotic education is an important curriculum to establish in schools to ensure that students are equipped to confront real-world issues since it allows teachers to produce talent that is ready for future skills. Collaboration and communication are important elements in making robotic education successful at the school level. STEM disciplines can be better adapted for robotics teaching. Apart from technical skills, soft skills can be achieved through robotic education, whether it is taught as a school curriculum subject or as informal robotic learning (extracurricular). The collaboration among teachers, parents, communities, industry, and government played essential roles in the success of robotic education in school.

References

1. Nagaraja, G. Dr., Shoba, H. Dr., Sreedevi, M. S. Dr., & Krishnamma, P. N. Dr. (2024). The impact of robotics and drones on agricultural efficiency and productivity. *International Journal of Research in Agronomy, 7*(9S), 1001–1009.
2. Yépez-Ponce, D. F., Salcedo, J. V., Rosero-Montalvo, P. D., & Sanchis, J. (2023). Mobile robotics in smart farming: Current trends and applications. *Frontiers in Artificial Intelligence, 6*, 1213330.

3. Yadav, A., Kholiya, D., Manjeet, R. B., Bhavana, J. B., Sah, B., & Bankoti, P. (2025). Agricultural robotics: Automating the future of farming. *Journal of Neonatal Surgery, 14*(6), 538–545.
4. Polonara, M., Romagnoli, A., Biancini, G., & Carbonari, L. (2024). Introduction of collaborative robotics in the production of automotive parts: A case study. *Machines, 12*(3), 196.
5. Huang, R., Li, H., Suomi, R., Li, C., & Peltoniemi, T. (2023). Intelligent physical robots in health care: Systematic literature review. *Journal of medical Internet research, 25*, Article e39786.
6. Kabanda, M. (2025). Artificial intelligence integration in higher education: Enhancing academic processes and leadership dynamics. *EIKI Journal of Effective Teaching Methods, 3*(1).
7. Ruano-Borbalan, J.-C. (2025). The transformative impact of artificial intelligence on higher education: A critical reflection on current trends and futures directions. *International Journal of Chinese Education, 14*(1).
8. Yusuf, A., Pervin, N., & Román-González, M. (2024). Generative AI and the future of higher education: A threat to academic integrity or reformation? Evidence from multicultural perspectives. *International Journal of Educational Technology in Higher Education, 21*, 21.
9. Sekaran, R. (2025, January 22). Pahang school makes history with victory at Int'l Robot Olympiad. *The Star*. https://www.thestar.com.my/news/nation/2025/01/22/pahang-school-makes-history-with-victory-at-int039l-robot-olympiad
10. Gratani, F., & Giannandrea, L. (2022), Towards 2030. Enhancing 21st century skills through educational robotics. *Frontiers in Education, 7*, 955285.
11. Ouyang, F., & Xu, W. (2024). The effects of educational robotics in STEM education: A multilevel meta-analysis. *International Journal of STEM Education, 11*, 7.
12. Bakan, G., & Bircan, M. A. (2025). Enhancing 21st century skills of primary school students in rural areas through STEM activities. *The Journal of Educational Research,* 1–16.
13. Avcı, F. (2024). Effects of robotic module-supported experimental activities with the cooperative learning method on student achievement, 21st-century skills, and students' opinions. *Journal of Computer Assisted Learning, 40*(5), 2325–2338.
14. Yilmaz, M. M., Bekirler, A., & Sigirtmac, A. D. (2024). Inspiring an early passion for science: The impact of hands-on activities on children's motivation. *ECNU Review of Education, 7*(4), 1033–1053.
15. Funk, M., Cascalho, J., Santos, A. I., Pedro, F., Medeiros, P., Amaral, B., Domingos, M., Ramos, A., & Mendes, A. (2022). A simple interactive robot to promote computational thinking. *Frontiers of Computer Science, 4*, 1022778.
16. Anwar, S., Bascou, N. A., Menekse, M., & Kardgar, A. (2019). A systematic review of studies on educational robotics. *Journal of Pre-College Engineering Education Research (J-PEER), 9*(2), Article 2.
17. Blanc, S., Conchado, A., Benlloch-Dualde, J. V., et al. (2025). Digital competence development in schools: A study on the association of problem-solving with autonomy and digital attitudes. *International Journal of STEM Education, 12*, 13.
18. Kaya-Capocci, S., Pabuccu-Akis, A., & Orhan-Ozteber, N. (2025). Entrepreneurial STEM education: Enhancing students' resourcefulness and problem-solving skills. *Research in Science Education, 55*, 103–134.
19. Ješková, Z., Lukáč, S., Šnajder, Ľ, Guniš, J., Klein, D., & Kireš, M. (2022). Active learning in STEM education with regard to the development of inquiry skills. *Education Sciences, 12*(10), 686.
20. Gravel, B. E., & Puckett, C. (2023). What shapes implementation of a school-based makerspace? Teachers as multilevel actors in STEM reforms. *International Journal of STEM Education, 10*, 7.
21. Kalaitzidou, M., & Pachidis, T. P. (2023). Recent robots in STEAM education. *Education Sciences, 13*(3), 272.

22. Sung, J., Lee, J. Y., & Chun, H. Y. (2023). Short-term effects of a classroom-based STEAM program using robotic kits on children in South Korea. *IJ STEM Ed, 10*, 26.
23. Khikmiyah, R., Rusijono, R., & Arianto, F. (2021). The effect of STEAM-robotics on science subjects on students' ability to solve problems. *AL-ISHLAH: Jurnal Pendidikan, 13*(3), 3061–3070.
24. Hsiao, H. S., Chen, J. H., Chang, T., et al. (2024). A study on the effects of using the 6E model and a robot teaching assistant on junior high school students' STEM knowledge, learning motivation, and hands-on performance. *Journal of Science Education and Technology, 33*, 759–778.
25. Bellas, F., Naya-Varela, M., Mallo, A., et al. (2024). Education in the AI era: A long-term classroom technology based on intelligent robotics. *Humanities and Social Sciences Communications, 11*, 1425.
26. Alam, A., & Mohanty, A. (2024). Integrated constructive robotics in education (ICRE) model: A paradigmatic framework for transformative learning in educational ecosystem. *Cogent Education, 11*(1), 2324487.
27. Shah, A. M., Wylie, C., Gitomer, D., & Noam, G. (2018). Improving STEM program quality in out-of-school-time: Tool development and validation. *Science Education, 102*, 238–259.
28. Coufal, P. (2022). Project-based STEM learning using educational robotics as the development of student problem-solving competence. *Mathematics, 10*(23), 4618.
29. Torres, A., Soliman, A., Isabell, T., et al. (2025). Teaching collaborative robotics: Design and evaluation of design-based learning curriculum for high school STEM education. *Journal of Formative Design in Learning, 9*, 77–91.

Core Principles of Robotics and Foundations of Programming

1 Introduction: The Convergence of Disciplines

Modern robotics emerges from the sophisticated integration of mechanical engineering, electrical systems, computer science, and control theory, creating complex cyber-physical systems that seamlessly interact with the physical world [1, 2]. This convergence has accelerated dramatically with the integration of artificial intelligence, cloud computing, and advanced sensor technologies, fundamentally transforming how we approach robotics programming and system design [3, 4]. The period from 2020 to 2025 has witnessed unprecedented advancement in AI-integrated robotics, collaborative systems, and edge computing implementations, necessitating a comprehensive understanding that bridges theoretical foundations with practical implementation considerations [5]. This evolution demands that robotics engineers and researchers master six critical, interconnected domains including hardware component understanding, programming language ecosystems, control systems theory, programming fundamentals, algorithmic development, and systematic debugging approaches.

2 Hardware Foundations and Programming Ecosystems

Modern robotic perception systems rely on sophisticated sensor architectures that bridge digital intelligence with physical environments. Vision sensors form the foundation, featuring RGB cameras with resolutions from 640 $\times$ 480 to 4 K at 30–240 fps and Time-of-Flight depth cameras achieving $\pm$ 1–5 mm accuracy over 0.1–10 m distances. Proximity sensing complements vision through ultrasonic sensors like the HC-SR04 (2 cm-4 m range with $\pm$ 3 mm theoretical accuracy, though practical performance varies), industrial sensors such as the Pepperl + Fuchs UC2000 (0.1–8 m range with IP67 rating), and LiDAR systems exemplified by the Velodyne

N. M. Thamrin et al., *Robotics in Education*,
SpringerBriefs in Applied Sciences and Technology,
https://doi.org/10.1007/978-981-95-3548-4_2

VLP-16 offering 360° horizontal and 30° vertical fields of view with 100 m range and 300,000 points per second data rates.

Computational processing integrates Inertial Measurement Units featuring high-precision gyroscopes (± 250°/s to ± 2000°/s range, 0.01°/s resolution), accelerometers (± 2 g to ± 16 g range with 16-bit resolution), and magnetometers (± 4 to ± 8 gauss range with 0.1° heading accuracy) through popular implementations like the Bosch BMI160 and InvenSense MPU-6050. Processing platforms span from Arduino's ATmega328P at 16 MHz for basic sensor integration to Raspberry Pi 4's quad-core ARM Cortex-A72 at 1.5 GHz for computer vision applications, while the NVIDIA Jetson Xavier NX delivers advanced AI capabilities with its 384-core Volta GPU, 48 Tensor Cores, and 6 Carmel ARM CPUs providing up to 21 TOPS of performance for real-time inference.

Actuation systems provide precise motion control through diverse motor technologies including brushed DC servos operating at 12–48 V with 0.1–100 Nm torque ranges, highly efficient brushless DC motors like the Maxon EC-max 30 delivering 134 W power with 339 mNm torque capacity, and stepper motors ranging from NEMA 17 units with 1.8° steps for lightweight applications to NEMA 34 systems for heavy-duty tasks. Linear actuators complement rotational motion through ball screw systems providing 10–1000 mm stroke with 0.01 mm resolution and 10–10,000N force capacity, while precision solutions like the Kollmorgen IC Series achieve 25–1200 mm stroke with 0.1 μm resolution for demanding manipulation requirements.

Recent innovations from 2020 to 2025 have transformed robotics capabilities through solid-state LiDAR technology exemplified by the Velodyne Velabit with 100 m range and no moving parts (originally targeting $100 price points in high-volume production), collaborative robotics systems implementing force limiting below 10 N for safe human–robot interaction, and advanced edge AI processors including the Google Coral Dev Board delivering 4 trillion operations per second at 0.5 watts per TOPS. These developments, combined with enhanced sensor fusion technologies and the Intel Movidius Myriad X for cloud-independent processing, collectively enable more autonomous, efficient, and safe robotic systems while democratising access to advanced robotics technologies across industrial, service, and research applications.

3 Programming Languages and Control Systems Integration

Python maintains its dominance in robotics programming through extensive AI and machine learning integration capabilities, supported by key libraries including NumPy for matrix operations and coordinate transformations, SciPy providing optimisation algorithms and signal processing capabilities essential for robotics applications, and OpenCV 4.11.0.86 delivering real-time image processing and object detection with native ROS sensor_msgs/Image message support. However, recent performance studies demonstrate Python consuming significantly more computational resources than C++ in ROS environments, with the Global Interpreter Lock (GIL) limiting true multithreading capabilities, though its rapid prototyping capabilities and extensive library ecosystem make it ideal for high-level control logic and AI integration where development speed outweighs execution performance [6].

ROS 2 represents a fundamental architectural evolution addressing many limitations of its predecessor through decentralised architecture, eliminating the central Master, DDS (Data Distribution Service) middleware providing quality-of-service guarantees necessary for real-time robotics applications, built-in security features with encrypted communication addressing cybersecurity concerns, and native multi-language support for C++ and Python APIs providing implementation flexibility. Communication paradigms in ROS 2 include asynchronous topics serving as the default mechanism for efficient data distribution, enhanced services providing improved error handling and timeout mechanisms for request-response operations, and actions combining topic and service functionality to support long-running operations with intermediate feedback capability.

C++ provides optimal performance for real-time robotics applications with substantially faster execution than Python for computational tasks, minimal memory overhead, and deterministic execution characteristics suitable for hard real-time constraints, while offering direct hardware access capabilities and extensive robotics libraries including OpenCV, PCL (Point Cloud Library), and Eigen making it essential for control systems and high-frequency sensor processing where performance is paramount. MATLAB and Simulink excel in control system design and simulation environments through rich toolboxes providing pre-implemented algorithms for robotics applications, graphical programming environments enabling rapid prototyping and algorithm validation, and enhanced ROS 2 integration through the ROS Toolbox, though licensing costs and deployment restrictions limit broader accessibility in commercial applications.

Emerging languages represent growing diversity in the robotics programming ecosystem, with Rust gaining traction for safety–critical systems through memory safety guarantees without garbage collection overhead, Julia showing promise for high-performance scientific computing applications where mathematical expressiveness and performance are both important, and Go demonstrating potential for concurrent programming and robotic service architectures particularly in cloud-connected and distributed robotics applications.

4 Control Systems and Mathematical Foundations

Proportional-Integral-Derivative (PID) control forms the mathematical cornerstone of robotics control systems through the fundamental control law $u(t) = K_p \cdot e(t) + K_i \cdot \int e(t)dt + K_d \cdot de(t)/dt$, where $u(t)$ represents the control output signal, $e(t)$ denotes the error signal calculated as setpoint minus measured value, K_p is the proportional gain coefficient determining immediate response to current error, K_i represents the integral gain coefficient eliminating steady-state errors through accumulated error integration, and K_d constitutes the derivative gain coefficient providing predictive action based on error rate of change [2, 7]. The discrete time implementation $u[k] = K_p \cdot e[k] + K_i \cdot \sum(e[i] \cdot \Delta t) + K_d \cdot (e[k] - e[k-1])/\Delta t$ enables digital control systems while introducing sampling considerations that directly impact control performance, where k represents the discrete time index and Δt denotes the sampling period requiring careful selection to maintain stability and performance.

State-space representation provides a systematic framework for modern multi-variable control design through the mathematical formulation $\dot{x}(t) = Ax(t) + Bu(t)$ and $y(t) = Cx(t) + Du(t)$, where $x(t)$ represents the state vector of dimension $n \times 1$ containing internal system variables, $u(t)$ denotes the input vector of dimension $m \times 1$ containing control signals, $y(t)$ constitutes the output vector of dimension $p \times 1$ containing measured variables, A represents the system matrix of dimension $n \times n$ characterising internal dynamics, B denotes the input matrix of dimension $n \times m$ relating inputs to state changes, C represents the output matrix of dimension $p \times n$ relating states to outputs, and D constitutes the feedthrough matrix of dimension $p \times m$ for direct input–output relationships [8]. This representation enables systematic analysis and control of multi-input, multi-output (MIMO) systems that characterise most robotic applications while facilitating optimal control design through Linear Quadratic Regulator (LQR) methodology.

Adaptive control systems address fundamental challenges of parameter uncertainty through Model Reference Adaptive Control (MRAC) employing the update law $\dot{\theta} = -\Gamma \cdot P^T \cdot e \cdot x$, where θ represents the adaptive parameter vector automatically adjusted in real-time, Γ denotes the adaptation gain matrix that must be positive definite to ensure stability, P constitutes the regressor matrix containing measured signals, e represents the tracking error between reference model and actual system response, and x denotes the state vector providing information for parameter adaptation. Self-tuning regulators represent an alternative adaptive approach combining online system identification with control design to provide automatic parameter adjustment for changing operating conditions, while robust control techniques including $H\infty$ control theory, optimise performance through minimising $\|T_{zw}\|\infty$ subject to internal stability constraints, providing guaranteed performance bounds even when plant parameters vary within specified uncertainty sets.

Manipulator control addresses the fundamental dynamic equation $M(q)\ddot{q} + C(q, \dot{q})\dot{q} + g(q) = \tau$ governing robot arm motion, where q represents the joint position vector of dimension $n \times 1$, $\dot{q}$ denotes the joint velocity vector of dimension $n \times 1$, $\ddot{q}$ constitutes the joint acceleration vector of dimension $n \times 1$, $M(q)$ represents the configuration-dependent inertia matrix of dimension $n \times n$ capturing mass and rotational inertia effects, $C(q, \dot{q})$ denotes the Coriolis and centrifugal force matrix of dimension $n \times n$ characterising velocity-dependent forces, $g(q)$ represents the gravitational force vector of dimension $n \times 1$ containing gravity effects, and τ constitutes the joint torque vector of dimension $n \times 1$ containing actuator commands [1]. Computed torque control implements model-based compensation through $\tau =$

$M(q)[\ddot{q}_d + K_v(\dot{q}_d - \dot{q}) + Kp(q_d - q)] + C(q, \dot{q})\dot{q} + g(q)$, where q_d, $\dot{q}_d$, and $\ddot{q}_d$ represent desired position, velocity, and acceleration trajectories, while K_p and K_v denote proportional and derivative gain matrices designed to achieve tracking performance specifications.

Kinematic control employs $\dot{q} = J^{(-1)(q)} \cdot \dot{x}_d$ using the Jacobian matrix $J(q)$ of dimension $6 \times n$ for 6-DOF end-effector motion relating joint velocities to end-effector velocity $\dot{x}_d$ of dimension 6×1, enabling intuitive control of robot motion in Cartesian coordinates while automatically handling complex coordinate transformations required for multi-joint systems [9]. Impedance control regulates interaction forces through $M(\ddot{x}_d - \ddot{x}) + B(\dot{x}_d - \dot{x}) + K(x_d - x) = F_{ext}$, where M, B, and K represent desired inertia, damping, and stiffness matrices, while F_{ext} denotes external forces, enabling compliant interaction essential for assembly and human–robot interaction applications.

5 Sensor Fusion and State Estimation

Kalman filtering provides optimal state estimation for linear systems through sophisticated mathematical frameworks balancing prediction and measurement information via prediction steps $\hat{x}^{(-)[k]} = A\hat{x}[k - 1] + Bu[k - 1]$ and $P^{(-)[k]} = AP[k - 1]A^T + Q$ projecting previous estimates forward based on system dynamics, where $\hat{x}^{(-)[k]}$ represents the a priori state estimate, $\hat{x}[k - 1]$ denotes the previous a posteriori state estimate, $P^{(-)[k]}$ constitutes the a priori error covariance matrix, $P[k - 1]$ represents the previous a posteriori error covariance matrix, and Q denotes the process noise covariance matrix characterising model uncertainty. Update steps employ Kalman gain $K[k] = P^{(-)[k]}H^T(HP^{(-)[k]}H^T + R)^{(-1)}$ and state correction $\hat{x}[k] = \hat{x}^{(-)[k]} + K[k](z[k] - H\hat{x}^{(-)[k]})$ with covariance update $P[k] = (I - K[k]H)P^{(-)[k]}$, where $z[k]$ represents the measurement vector, H denotes the measurement matrix relating states to measurements, R constitutes the measurement noise covariance matrix, and I represents the identity matrix.

The Extended Kalman Filter (EKF) extends this framework to nonlinear systems through linearisation $F = \partial f/\partial x|\{\hat{x}, u\}$ and $H = \partial h/\partial x|\{\hat{x}^{(-)}\}$, where F represents the Jacobian of the nonlinear process function $f(\cdot)$ with respect to states, H denotes the Jacobian of the nonlinear measurement function $h(\cdot)$ with respect to states, enabling application to the nonlinear systems characterising most robotics applications. Recent advances from 2020 to 2025 include distributionally robust Kalman filters addressing uncertainty in noise distributions using Wasserstein ambiguity sets providing robust performance when noise characteristics are poorly known [10], and adaptive extended Kalman filters incorporating real-time noise parameter estimation, enabling performance optimisation as operating conditions change [11].

Particle filters employ sequential Monte Carlo methods representing posterior distributions through particle sets $p(x_k|z_{\{1:k\}}) \approx \Sigma\, w_k^{(i)} \cdot \delta(x_k - x_k^{(i)})$, where $w_k^{(i)}$ represents the weight of particle i at time k, $\delta(\cdot)$ denotes the Dirac Delta function, and $x_k^{(i)}$ constitutes the state of particle i, enabling handling of highly nonlinear systems and non-Gaussian uncertainty that cannot be adequately addressed through Kalman filtering approaches. Applications include robot localisation in complex environments where linear assumptions break down, SLAM in dynamic environments with moving obstacles, and multi-target tracking scenarios where the number of targets varies over time.

6 Algorithm Development and Advanced Navigation

A* path planning represents the optimal fusion of systematic search with intelligent heuristic guidance through the cost function $f(n) = g(n) + h(n)$, where $f(n)$ represents the total estimated cost from start to goal through node n, $g(n)$ denotes the actual cost from start to node n accumulated along the path, and $h(n)$ constitutes the heuristic estimate from node n to goal that must satisfy the admissibility condition $h(n) \leq h^*(n)$ where $h^*(n)$ represents the true optimal cost to ensure optimal solutions [12]. The algorithm's effectiveness stems from focusing search efforts on promising regions while maintaining mathematical guarantees of optimality, though performance characteristics include exponential complexity $O(b^d)$ in worst-case scenarios where b represents branching factor and d denotes solution depth.

Rapidly-exploring Random Trees (RRT) provide probabilistically complete solutions for high-dimensional planning problems through iterative sampling of random configurations x_rand, finding nearest existing nodes x_near, and extending the tree toward sampled points by step size δ, creating data structures that rapidly explore configuration space while maintaining connectivity for path reconstruction [12]. RRT* optimisation introduces asymptotic optimality through sophisticated rewiring mechanisms with radius $r^* = \gamma(\log(n)/n)^{(1/d)}$, where γ represents a scaling constant, n denotes the number of nodes in the tree, and d constitutes the dimensionality of the configuration space, achieving cost function convergence as iterations increase and ensuring solution quality improves continuously with additional computation [13]. Recent advances demonstrate significant performance improvements through hybrid approaches combining algorithm strengths, with A*-RRT* hybrid algorithms achieving substantial speed improvements over traditional RRT-Connect while reducing average path length [14], and Bidirectional Informed-RRT* (BI-RRT*) reducing CPU running time by limiting the sampling space of random points against Informed-RRT* [15, 16].

Dynamic Window Approach (DWA) operates directly in velocity space considering robot dynamics for real-time obstacle avoidance through the dynamic window $V_d = \{(v,\omega) \mid v \in [v_{min}, v_{max}], \omega \in [\omega_{min}, \omega_{max}]\}$ representing achievable velocities based on current robot state and actuator capabilities, while admissible velocities $V_a = \{(v,\omega) \mid v \leq \sqrt{(2d \cdot a)}, \omega \leq \sqrt{(2\theta \cdot \alpha)}\}$ ensure collision-free stopping considering robot dynamics and obstacle proximity, where v and ω represent linear and angular velocities, d denotes distance to nearest obstacle, θ represents angular distance to obstacle, and a and α constitute maximum linear and angular accelerations [17]. The objective function $G(v,\omega) = \sigma(\alpha \cdot \text{heading}(v,\omega) + \beta \cdot \text{dist}(v,\omega) + \gamma \cdot \text{vel}(v,\omega))$ optimises multiple criteria including heading toward goal, obstacle clearance, and velocity preferences through weighted combination using normalisation function σ and weighting coefficients α, β, and γ enabling tuning of robot behaviour for different applications.

Simultaneous Localisation and Mapping (SLAM) addresses fundamental challenges through Extended Kalman Filter approaches maintaining joint estimation of robot pose and landmark positions via state vector $x = [x_r, y_r, \theta_r, x_1, y_1, x_2, y_2, \ldots, x_n, y_n]^T$ that grows with observed landmarks, where (x_r, y_r, θ_r) represents robot pose including position and orientation, (x_i, y_i) denotes landmark i position coordinates, and n constitutes the number of observed landmarks [18, 19]. Graph-based SLAM formulates problems as optimisation, minimising chi-squared error $\chi^2 = \Sigma\ e_{\{ij\}}{}^T\ \Omega_{\{ij\}}\ e_{\{ij\}}$, where $e_{\{ij\}}$ represents the error between poses i and j, $\Omega_{\{ij\}}$ denotes the information matrix representing inverse covariance, and summation occurs over all pose-pose and pose-landmark constraints enabling efficient solutions for large-scale mapping through sparse matrix techniques.

Visual SLAM systems employ feature-based approaches using distinctive visual features, including SIFT, SURF, or ORB, with matching and tracking algorithms maintaining correspondence across image sequences, while direct methods minimise photometric errors for dense reconstruction using all pixel information rather than sparse features. Recent developments include ORB-SLAM3 providing visual-inertial SLAM capabilities with real-time performance combining camera and IMU data for improved robustness, and neural SLAM approaches using deep learning for feature extraction and loop closure detection that adapt to new environments through experience [20].

7 Debugging and System Validation

Robotic system reliability is achieved through structured debugging and validation methodologies that span the entire development lifecycle. The V-Model provides a systematic framework linking each development phase, i.e., system requirements, high-level design, detailed design, and implementation with corresponding verification phases, namely, unit, integration, system, and acceptance testing [21]. To enhance robustness, Corroborative Verification and Validation (V&V) integrates multiple validation approaches. Formal verification ensures correctness using model checking and temporal logic, while simulation-based testing emulates physical environments to explore edge cases. Real-world validation incorporates hardware-in-the-loop and staged field testing to capture complex operational conditions [22].

Root cause analysis underpins fault isolation and correction by identifying fundamental issues beyond surface symptoms. Techniques such as the “5 Whys” and divide-and-conquer isolate faulty subsystems, supported by data collection and simulation validation before deployment. Debugging the hardware-software interface involves signal integrity checks, protocol compliance, and voltage compatibility analysis to resolve issues like communication timeouts or erratic sensor outputs.

Timing constraint debugging addresses temporal reliability through tools that analyse scheduling policies, detect resource contention, and profile execution time. These help diagnose issues such as control loop overruns and missed deadlines. Modern robotics also benefit from advanced debugging tools. ROS 2 enhancements provide richer logging and diagnostic data, with visualisation platforms like Foxglove Studio, RViz2, and PlotJuggler supporting real-time state monitoring. AI-assisted debugging represents a frontier, applying machine learning for predictive maintenance and anomaly detection in sensor data.

8 Educational Excellence and Future Directions

Leading educational institutions demonstrate diverse but complementary approaches to robotics education, with MIT exemplifying integration of theoretical rigour with hands-on implementation through 2.12 Introduction to Robotics, emphasising systematic progression from planar and spatial kinematics through motion planning, mechanism design, multi-rigid-body dynamics, and 3D graphic simulation supported by weekly laboratory sessions providing direct experience with servo drives, real-time control, and embedded software implementation. Stanford University provides systematic coverage through CS223A Introduction to Robotics encompassing spatial transformations, forward and inverse kinematics, Jacobians, robot dynamics, and multiple control methodologies with assessment frameworks employing homework assignments, examinations, and critical thinking methodologies preparing students for advanced research and applications.

Carnegie Mellon University operates the most extensive university-based robotics education initiative through the Carnegie Mellon Robotics Academy (CMRA) functioning as research, development, and outreach program investigating how buildable and programmable agents teach CS-STEM across educational levels, with the foundational Mobile Robot Programming Course employing problem-driven approaches for students with minimal robotics experience encompassing perception, action, and cognition while advancing to sophisticated topics including reinforcement learning and multi-robot coordination. The CS-STEM Network (CS2N) provides comprehensive online tools and curricula with fully virtual solutions including programming environments and virtual robots embedded within curriculum materials, enabling teachers to earn Carnegie Mellon Robotics Academy Certification with continuing education credits.

Multidisciplinary integration exemplified by Georgia Institute of Technology's Institute for Robotics and Intelligent Machines (IRIM) operates the first truly multidisciplinary robotics Ph.D. program jointly offered by Computing and Engineering colleges emphasising comprehensive coverage of five core areas including Mechanics, Control, Perception, Artificial Intelligence, and Human–Robot Interaction, while University of Pennsylvania administers unique Master of Science in Engineering (MSE) in Robotics through the GRASP Laboratory with multidisciplinary programs sponsored by Computer and Information Science, Electrical and

Systems Engineering, and Mechanical Engineering departments requiring courses in at least three foundational areas.

A multilevel meta-analysis was presented by [23] to investigate the effectiveness of educational robotics in STEM education across 21 studies and 30 effect sizes involving 2433 participants. The findings indicate that educational robotics have a moderate overall positive effect (Hedges' $g = 0.488$), particularly enhancing students' learning performance ($g = 0.665$) and learning attitudes ($g = 0.497$), but with no significant impact on computational thinking ($g = 0.079$). Among eight moderator variables analysed, only the STEM discipline significantly influenced the outcomes, with technology and cross-disciplinary applications showing stronger effects than science and mathematics. Other variables, such as educational level, instructor support, instructional strategy, interaction type, intervention duration, robotic type, and control group condition, showed varying trends but were not statistically significant. The study highlights the pedagogical importance of aligning robotic interventions with discipline-specific goals and emphasises the role of instructors and student-centred strategies for maximising the benefits of robot-assisted STEM education.

Emerging technological paradigms include quantum computing applications in robotics optimisation, representing frontiers that could revolutionise path planning and resource allocation problems through quantum algorithms showing particular promise for solving complex optimisation problems computationally intractable for classical computers, potentially enabling real-time optimisation of multi-robot coordination and resource allocation previously impossible. Neuromorphic computing offers revolutionary approaches to energy-efficient processing mimicking brain-like computation models excelling at pattern recognition, adaptive learning, and parallel processing tasks central to robotics applications while consuming orders of magnitude less power than traditional digital processors [24].

Bio-inspired control algorithms draw from biological systems creating more adaptive, robust, and efficient control strategies enabling robots to exhibit behaviours more closely approximating biological intelligence including adaptive learning, emergent behaviours, and robust operation in unpredictable environments [25]. Autonomous debugging systems represent next frontiers in robotics reliability employing AI-driven approaches to automatically detect, diagnose, and potentially repair system problems without human intervention through combining anomaly detection, predictive analytics, and automated response mechanisms maintaining system operation even when individual components fail [26].

Digital twin technologies create virtual replicas of physical robotic systems enabling sophisticated simulation, prediction, and optimisation capabilities through predictive maintenance, performance optimisation, and safety validation in virtual environments before deployment to physical systems [27]. Self-adapting robotic systems employ machine learning and adaptive control techniques continuously improving performance based on experience and changing environmental conditions, representing fundamental shifts from static pre-programmed behaviours toward dynamic learning-based adaptation, enabling robots to operate effectively in environments and situations not explicitly anticipated during design [28].

9 Conclusion: Integration and Future Perspectives

Modern robotics programming represents sophisticated integration of theoretical foundations with practical implementations, driven by AI enhancement, advanced computational methods, and robust debugging methodologies creating unprecedented opportunities for autonomous systems extending beyond traditional industrial applications [1, 3]. The convergence of hardware miniaturisation, AI-enhanced algorithms, and comprehensive software frameworks enables new classes of applications while requiring new approaches to education, development, and deployment balancing innovation with reliability, capability with safety, and performance with maintainability.

Key findings demonstrate successful robotics programming requires multidisciplinary expertise spanning hardware understanding, multiple programming languages, control theory, algorithmic design, and systematic debugging approaches, with integration through frameworks like ROS 2 combined with AI capabilities and cloud-edge computing representing current state-of-the-art development practices [29]. The evolution toward adaptive learning-based systems reflects fundamental transformation from simple pre-programmed behaviours to sophisticated autonomous intelligence requiring comprehensive understanding of theoretical principles and practical implementation challenges combined with systematic debugging and validation approaches for reliable operation in complex dynamic environments.

Future research directions encompass quantum computing applications in robotics optimisation, potentially revolutionising computational approaches, neuromorphic computing for energy-efficient processing enabling new classes of applications, bio-inspired control algorithms creating more adaptive and robust behaviours, and autonomous debugging systems maintaining operation without human intervention. The integration of digital twin technologies, predictive maintenance capabilities, and self-adapting robotic systems represents significant opportunities for advancing capabilities while addressing reliability and maintenance challenges currently limiting widespread deployment.

As robots become increasingly prevalent in society, these fundamental principles will continue serving as foundations for more sophisticated autonomous systems, with convergence of artificial intelligence, advanced computing architectures, and sensor technologies promising continued innovation expanding robotic applications across diverse domains from healthcare and environmental stewardship to space exploration and disaster response. Success in robotics programming increasingly requires not just technical competence but systematic thinking, collaborative development skills, and ability to integrate knowledge across multiple disciplines, with practitioners mastering these diverse requirements driving next-generation innovations extending robotic capabilities into new domains while addressing grand challenges facing society.

References

1. Siciliano, B., Sciavicco, L., Villani, L., & Oriolo, G. (2022). *Robotics: Modelling, planning and control* (2nd ed.). Springer.
2. Spong, M. W., Hutchinson, S., & Vidyasagar, M. (2020). *Robot modeling and control* (2nd ed.). Wiley.
3. Billard, A., Albu-Schaeffer, A., Beetz, M., Burgard, W., Corke, P., Ciocarlie, M., Dahiya, R., Kragic, D., Goldberg, K., Nagai, Y., & Scaramuzza, D. (2025). A roadmap for AI in robotics. *Nature Machine Intelligence*, 1–7.
4. Russell, S., & Norvig, P. (2021). *Artificial intelligence: A modern approach* (4th ed.). Pearson.
5. Macenski, S., Foote, T., Gerkey, B., Lalancette, C., & Woodall, W. (2022). Robot operating system 2: Design, architecture, and uses in the wild. *Science Robotics, 7*(66), Article eabm6074. https://doi.org/10.1126/scirobotics.abm6074
6. Albonico, M., Cannizza, M. B., & Wortmann, A. (2025). Energy efficiency in ROS communication: A comparison across programming languages and workloads. *Frontiers in Robotics and AI, 12*, 1548250.
7. Corke, P. (2017). *Robotics, vision and control: Fundamental algorithms in MATLAB* (2nd ed.). Springer.
8. Lynch, K. M., & Park, F. C. (2017). *Modern robotics: Mechanics, planning, and control.* Cambridge University Press.
9. Niku, S. B. (2020). *Introduction to robotics: Analysis, control, applications* (3rd ed.). Wiley.
10. Jang, M., Hakobyan, A., & Yang, I. (2025). On the steady-state distributionally robust Kalman filter. arXiv preprint arXiv:2503.23742. https://arxiv.org/abs/2503.23742
11. Kou, E., & Haggenmiller, A. (2023). Extended Kalman filter state estimation for autonomous competition robots. arXiv preprint arXiv:2310.04459. https://arxiv.org/abs/2310.04459
12. LaValle, S. M. (2006). *Planning algorithms.* Cambridge University Press.
13. Karaman, S., & Frazzoli, E. (2011). Sampling-based algorithms for optimal motion planning. *International Journal of Robotics Research, 30*(7), 846–894. https://doi.org/10.1177/0278364911406761
14. Zhang, R., Guo, H., Andriukaitis, D., Li, Y., Królczyk, G., & Li, Z. (2024). Intelligent path planning by an improved RRT algorithm with dual grid map. *Alexandria Engineering Journal, 88*, 91–104.
15. Fan, H., Huang, J., Huang, X., Zhu, H., & Su, H. (2024). BI-RRT*: An improved path planning algorithm for secure and trustworthy mobile robots systems. *Heliyon, 10*(5).
16. Gammell, J. D., Srinivasa, S. S., & Barfoot, T. D. (2014). Informed RRT*: Optimal sampling-based path planning focused via direct sampling of an admissible ellipsoidal heuristic. In *2014 IEEE/RSJ international conference on intelligent robots and systems* (pp. 2997–3004). https://doi.org/10.1109/IROS.2014.6942976
17. Fox, D., Burgard, W., & Thrun, S. (1997). The dynamic window approach to collision avoidance. *IEEE Robotics & Automation Magazine, 4*(1), 23–33. https://doi.org/10.1109/100.580977
18. Durrant-Whyte, H., & Bailey, T. (2006). Simultaneous localization and mapping: Part I. *IEEE Robotics & Automation Magazine, 13*(2), 99–110. https://doi.org/10.1109/MRA.2006.1638022
19. Thrun, S., Burgard, W., & Fox, D. (2005). *Probabilistic robotics.* MIT Press.
20. Zhang, Y., Wu, Y., Tong, K., Chen, H., & Yuan, Y. (2023). Review of visual simultaneous localization and mapping based on deep learning. *Remote Sensing, 15*(11), 2740.
21. Luckcuck, M., Farrell, M., Dennis, L. A., Dixon, C., & Fisher, M. (2019). Formal specification and verification of autonomous robotic systems: A survey. *ACM Computing Surveys, 52*(5), 1–41. https://doi.org/10.1145/3342355
22. Webster, M., Western, D., Araiza-Illan, D., Dixon, C., Eder, K., Fisher, M., & Pipe, A. G. (2020). A corroborative approach to verification and validation of human-robot teams. *International Journal of Robotics Research, 39*(1), 73–99. https://doi.org/10.1177/0278364919883338

23. Ouyang, F., & Xu, W. (2024). The effects of educational robotics in STEM education: A multilevel meta-analysis. *International Journal of STEM education, 11*(1), 7. https://doi.org/10.1186/s40594-024-00469-4
24. Putra, R. V. W., Wickramasinghe, P., & Shafique, M. (2025). Enabling efficient processing of spiking neural networks with on-chip learning on commodity neuromorphic processors for edge AI systems. arXiv preprint arXiv:2504.00957
25. Patanè, L., & Zhao., G. (2023) Editorial: Focus on methods: neural algorithms for bio-inspired robotics. *Frontiers in Neurorobotics, 17*, 1250645. https://doi.org/10.3389/fnbot.2023.1250645
26. Chirayil Nandakumar, S., Mitchell, D., Erden, M. S., Flynn, D., & Lim, T. (2024). Anomaly detection methods in autonomous robotic missions. *Sensors, 24*(4), 1330.
27. Mourtzis, D., Tsoubou, S., & Angelopoulos, J. (2023). Robotic cell reliability optimization based on digital twin and predictive maintenance. *Electronics, 12*(9), 1999.
28. Rezende Silva, G., Päßler, J., Tapia Tarifa, S. L., Johnsen, E. B., & Hernández Corbato, C. (2025). ROSA: A knowledge-based solution for robot self-adaptation. *Frontiers in Robotics and AI, 12*, 1531743.
29. Araujo, H., Mousavi, M. R., & Varshosaz, M. (2023). Testing, validation, and verification of robotic and autonomous systems: A systematic review. *ACM Transactions on Software Engineering and Methodology, 32*(2), 1–61.

Mechatronics Design in Robotics: Mechanical, Control and Sensor Systems Integration

1 Introduction

Robotics education hinges on fundamentals of robot system itself in which the integration of mechanical, electrical, and sensory systems to create intelligent, functional machines as shown in Fig. 1. Understanding this synergy is crucial for students aiming to develop robust robotic platforms capable of operating in diverse environments. As depicted in Fig. 1, robotics falls under the category of controlled, man-made mechanical systems. This categorisation underscores the necessity of integrating mechanical components, control logic, and sensor feedback into a unified framework.

The focus of this chapter is to guide students through the core elements of mechatronic design in robotics, including mechanism development, hierarchical control strategies, sensor integration, and real-world implementation. Starting from basic remote-controlled systems and advancing toward fully autonomous robots, this structured approach provides a practical foundation for mastering system integration in robotics.

2 Understanding Mechanism Design in Educational Robotics

2.1 Robot Mechanisms and Configuration

Understanding robot mechanisms is essential in robotics education, enabling students to design systems suited to specific tasks and environments. Figure 2 shows various example of robot configurations for operation in water, land, air, and hybrid mediums. Each configuration presents unique engineering challenges and performance advantages.

N. M. Thamrin et al., *Robotics in Education*,
SpringerBriefs in Applied Sciences and Technology,
https://doi.org/10.1007/978-981-95-3548-4_3

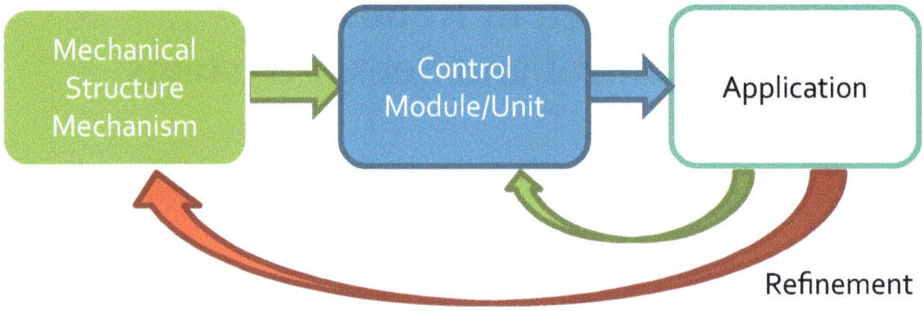

Fig. 1 Basic robotics systems

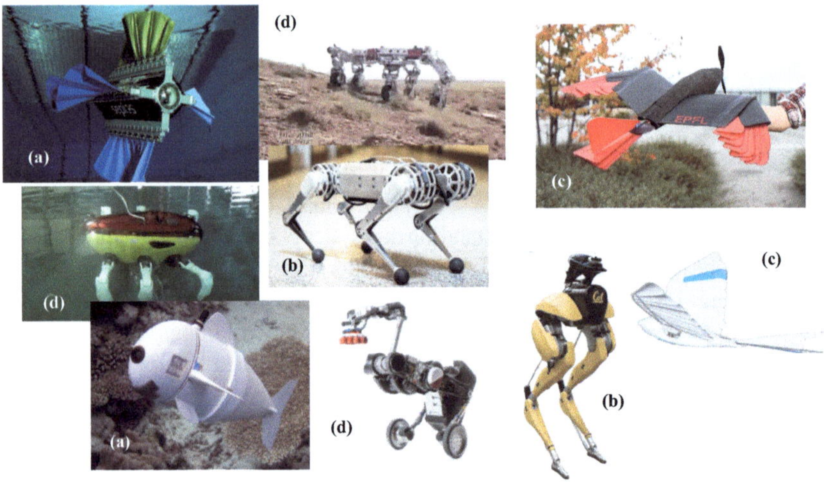

Fig. 2 Example of mobile robot mechanisms for **a** water, **b** ground, **c** air and **d** hybrid mediums

Underwater robots often use fin- or leg-based mechanisms as depicted in Fig. 2a [1, 2], allowing mobility in submerged environments both swimming and testbed navigations. These systems are used for marine exploration and inspection but must overcome high pressure and limited communication. On land environment, Legged robots as shown in Fig. 2b [3, 4], offer superior terrain adaptability and are commonly applied in search-and-rescue and planetary missions. Their performance depends on complex kinematic models and advanced control strategies.

For aerial operations, robots designed with fixed-wing or flapping-wing mechanisms, as shown in Fig. 2c [5, 6], are preferred for fast coverage for surveillance and delivery tasks, though they are constrained by short battery life and weather sensitivity. Conversely, wheeled and hybrid robots perform best on flat surfaces, multi-tasking environment and offer unity in control unit design. Many modern platforms integrate wheels, legs, or manipulators for multi-environment capability. For example, hybrid systems can transition between aerial and underwater operations, as shown in Fig. 2d [7–9].

Fundamentally, robot configuration refers to the arrangement of joints, links, and actuators that define motion capabilities and workspace. Selecting an appropriate configuration requires balancing precision, reach, and environmental adaptability. The Cartesian configuration with Prismatic-Prismatic-Prismatic (PPP) configuration uses three linear joints aligned along orthogonal axes as depicted in Fig. 3a and is ideal for high-precision applications like CNC machining and 3D printing. However, it demands significant space and is less durable in harsh conditions. Therefore, Cylindrical robot configurations with Revolute-Prismatic-Prismatic (RPP) as shown in Fig. 3(b) [10] was introduced to overcome vertical reach in compact spaces. Moreover, Spherical or polar robots (RRP configuration) allow hemispherical movement as shown in Fig. 3(c) [10] are preferred for wide lateral motions such as welding and spray painting. With the advancement of technology, the Selective Compliance Assembly Robot Arm (SCARA), using either RRP or PRR configurations, excels in fast horizontal tasks such as electronic assembly. Its selective compliance enhances alignment during insertion tasks, as illustrated in Fig. 3d [11].

More advance robot technology come out with articulated configuration with multiple rotary joints resemble human arms as shown in Fig. 3e [12]. This robot offering high dexterity for obstacle navigation. They are widely used in industrial applications such as welding and packaging, exemplified by ABB's IRB 6700 and Universal Robots' UR10e.

Conversely parallel manipulator was introduced name as Delta robot (Fig. 3f [13]), achieve exceptional speed and precision using three lightweight arms connected to

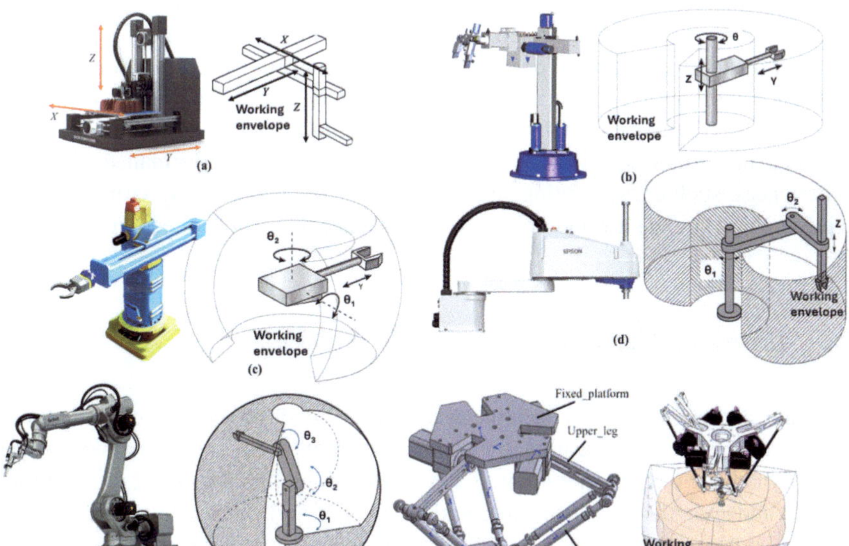

Fig. 3 Example of common robot mechanism and configurations with working envelope

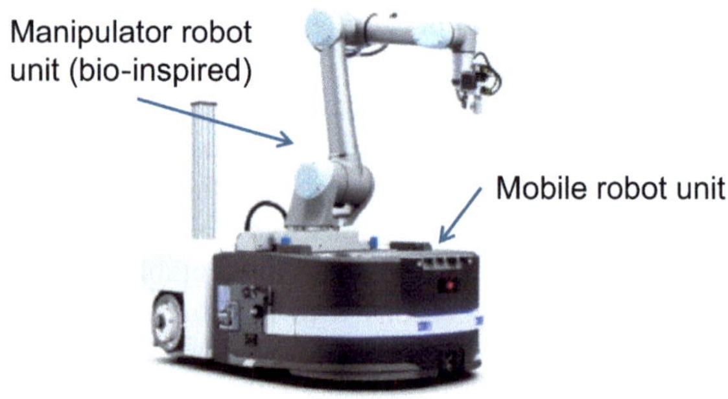

Fig. 4 Example of hybrid and complex robots

a shared end-effector. While ideal for high-speed pick-and-place operations, their dome-shaped workspace limits payload capacity. Beyond conventional structures, hybrid and bio-inspired configurations expand robot functionality through novel combinations of kinematic elements. As illustrated in Fig. 4, mobile manipulators integrate wheeled bases with articulated arms for dynamic environments. Bio-inspired mechanisms emphasise adaptability and energy efficiency, though they face limitations in payload and computational demand [14].

Choosing the right mechanism and configuration hinges on task-specific requirements, including precision, speed, environmental constraints, payload, and cost. Cartesian and SCARA robots suit structured environments requiring repeatability, while articulated configurations offer unmatched dexterity in complex or confined spaces. Delta robots are ideal for high-speed light-duty operations, and bio-inspired systems thrive in unpredictable terrains due to their inherent adaptability.

In summary, the structural design of robot mechanisms underpins their mobility and task capabilities. Familiarity with configurations such as Cartesian, cylindrical, spherical, SCARA, articulated, Delta, hybrid, and bio-inspired forms allows students to understand trade-offs between reach, complexity, and performance. Practical experiences, such as assembling simple 3-axis robots or simulating virtual models, reinforce this knowledge and prepare learners for real-world application.

2.2 *Basics of Mechanical Design for Students*

Mechanical design serves as the structural foundation of robotic systems and plays a critical role in robotics education. It introduces students to essential principles such as force dynamics, torque generation, material behaviour, and the use of design tools. These fundamentals are rooted in classical physics and further enhanced through modern simulation technologies, offering students a comprehensive platform to develop practical engineering competence [15].

At the foundation of robotic mechanics is Newton's second law, expressed as:

$$F = ma \tag{1}$$

This equation helps determine the force required for a given motion and is crucial when selecting actuators. For instance, lifting a 2 kg payload with an acceleration of 3 m/s^2 requires at least 6 N of upward force. When rotational motion is involved, torque becomes the dominant factor. Torque (τ) is defined by the formula:

$$\tau = r \times F \tag{2}$$

where r represents the distance from the pivot point. Torque significantly affects the performance of robotic joints and manipulators, especially in arm-type structures with revolute joints. In addition to understanding motion forces, students must also grasp how materials respond to mechanical loads. Stress, which refers to force per unit area, and strain, which measures the relative change in length, are described by the following equations:

$$\sigma = \frac{F}{A}, \varepsilon = \frac{\Delta L}{L_0} \tag{3}$$

where

- σ: stress (force per unit area)
- ε: strain (relative change in length)
- A: cross-sectional area
- ΔL: change in length
- L_0: original or initial length.

These values guide students in selecting appropriate materials based on load-bearing capacity and deformation limits. For example, ABS plastic can withstand up to 40 MPa before deforming, whereas aluminium can tolerate stresses exceeding 200 MPa. Understanding these properties enables students to make informed decisions about material selection for different robot components, balancing strength, weight, and flexibility. Figure 5a illustrates how forces and torque distribute across a robotic arm, while Fig. 5b shows a typical stress–strain curve, highlighting both elastic and plastic deformation under load.

To apply these theoretical concepts effectively, students increasingly rely on computer-aided design (CAD) software. Entry-level tools like TinkerCAD allow beginners to create basic models using geometric primitives. More advanced platforms such as Fusion 360 and SolidWorks offer features including parametric modelling, motion simulation, and stress analysis. These tools help students visualise how forces act on their designs, identify weak points, and refine dimensions for better performance. Simulation-based workflows significantly enhance learning by enabling virtual testing and refinement before building physical prototypes. For example, a design course at Stanford University reported a 60% reduction in prototyping costs after students used Fusion 360 to simulate drone landings and correct

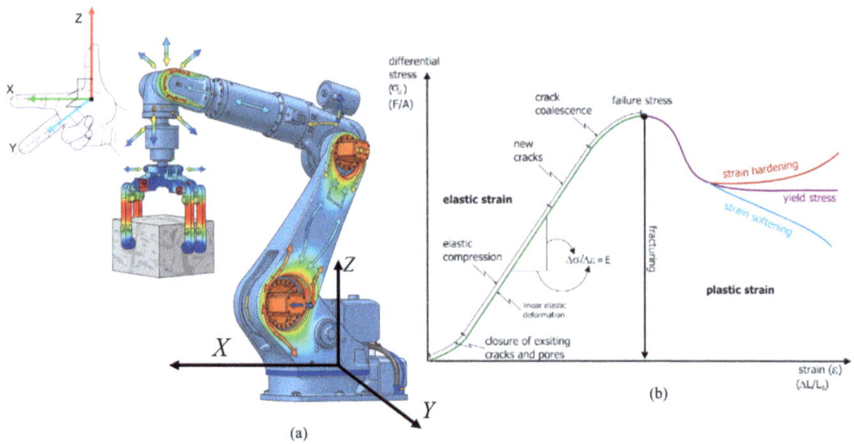

Fig. 5 Illustration depicting a robotic arm; **a** lifting an object, highlighting areas experiencing force, torque, stress, and strain. **b** Example of Stress–strain graph showing material behaviour under loading, illustrating elastic and plastic strain regions

structural weaknesses early in the design process [16]. These examples demonstrate how digital tools reinforce engineering principles and promote analytical thinking.

Integrating mechanical theory with CAD and simulation tools not only builds technical proficiency but also deepens students' understanding of structural behaviour. By engaging with both physical principles and digital environments, learners gain the skills needed to design robots that are functional, durable, and optimised for performance [17].

2.3 *Pedagogical Strategies and Practical Implementations*

Effective teaching of mechanical design in robotics benefits greatly from structured, hands-on learning approaches. Among these, project-based learning (PBL) has proven particularly impactful, enabling students to apply mechanical principles through progressive stages of design, construction, and testing. By engaging in real-world tasks, students not only deepen their understanding of force, torque, and material behaviour but also develop critical thinking and engineering design skills.

A typical PBL framework in robotics begins with structure assembly, where students use kits such as LEGO Mindstorms or VEX Robotics to build foundational elements like chassis, arms, or frames. This stage emphasises basic mechanics, allowing students to explore how geometry and load paths affect force distribution and joint performance. Once structural stability is achieved, students move on to mechanism integration, incorporating motors, gears, and sensors to transform static models into functional robots. Tasks often include simple behaviours such as line following, obstacle avoidance, or object manipulation. The final stage involves design optimisation, where students test their robots and iterate based on performance data. They may adjust gear ratios to increase speed, redistribute weight to improve balance, or modify linkages to enhance efficiency. This iterative process mirrors real engineering workflows and reinforces the importance of evaluation and continuous improvement. Research indicates that students engaged in such PBL activities consistently outperform peers in theoretical-only learning environments, both in comprehension and retention of mechanical concepts [18].

Emerging technologies further enrich these learning experiences. Augmented Reality (AR), for example, allows students to visualise internal forces, stress concentrations, and torque distributions directly overlaid on physical models using wearable headsets (see Fig. 6). This immersive visualisation helps bridge the gap between abstract theory and real-time structural behaviour.

In parallel, digital twins offer virtual counterparts of robotic systems that respond to input parameters such as motor torque or material stiffness. Students can simulate and compare outcomes without risking hardware damage, fostering experimentation and safe failure analysis.

Furthermore, AI-powered design assistants are increasingly integrated into educational platforms. These tools analyse robot tasks and provide design suggestions, such as reshaping a gripper to improve contact stability or reducing material usage for lighter weight. Such enhancements not only make learning more engaging but also expose students to the intelligent design support systems used in modern engineering.

In summary, combining project-based learning with advanced visualisation and simulation tools significantly enhances mechanical design education. Students gain both theoretical knowledge and practical experience, equipping them with

Fig. 6 An AR headset user examining a robotic arm with overlaid colour maps showing torque and stress distribution

the creativity and problem-solving skills necessary for successful robotic system development.

3 Control Systems and Electronics Design in Educational Robotics

Control systems are central to robotic functionality, acting as the bridge between mechanical structures, sensors, and operational goals. In educational robotics, mastering control principles enables students to design systems that are responsive, stable, and capable of adapting to real-world conditions. As shown in Fig. 7, a closed-loop system integrates sensors, controllers, and actuators through continuous feedback to maintain desired behaviour [19, 20].

3.1 Core Concepts in Control and Programming

Effective control system design in robotics revolves around four key objectives: stability, accuracy, responsiveness, and robustness. Stability ensures consistent robot behaviour despite external disturbances, forming a foundational concept in both theoretical frameworks and practical projects [21–23]. Accuracy focuses on minimising errors during steady-state and transient phases, often evaluated using metrics such as root mean square error (RMSE) and mean absolute error (MAE) [24]. Responsiveness refers to how quickly a robot reacts to inputs, while robustness measures the system's ability to maintain performance under uncertain or changing conditions, typically addressed through adaptive or robust control strategies [25].

Before diving into programming, it is essential to understand the robot's hierarchical control architecture (HCA), which organises the control system into modular components for efficient operation. As illustrated in Fig. 8, the HCA typically consists of three primary modules that operate in a coordinated manner to ensure accurate

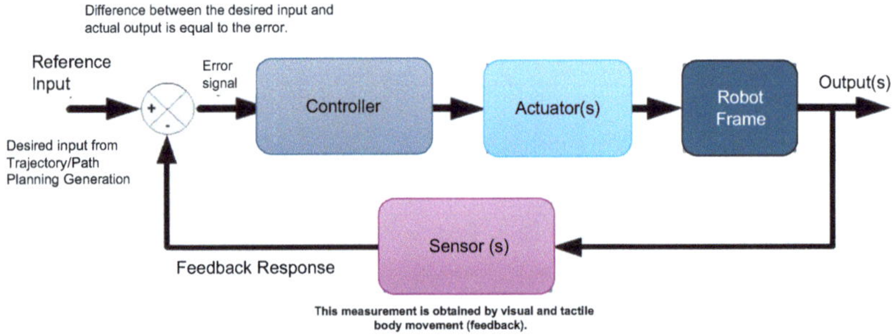

Fig. 7 General closed-loop robot system design

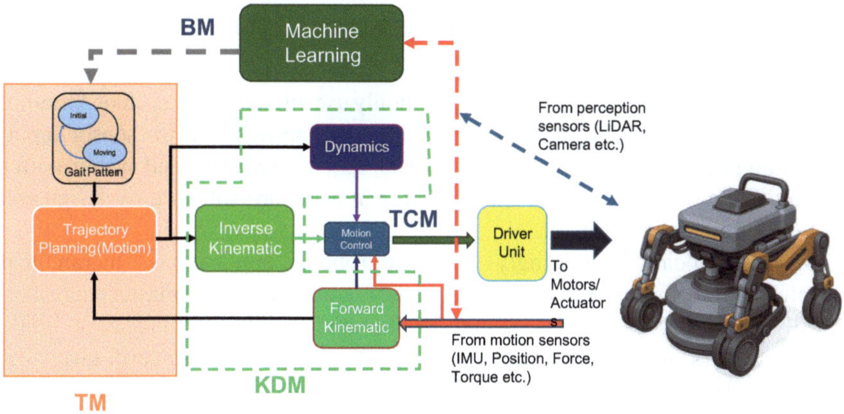

Fig. 8 Example of HCA for an autonomous mobile robot

robotic motion. Initially, the Trajectory Module (TM), situated within the master controller, is responsible for generating the reference path or trajectory that the robot is intended to follow. Subsequently, the Kinematics and Dynamics Module (KDM) interpret this trajectory and converts it into specific motion commands suitable for the robot's drive units (commonly voltage signal). Thereafter, the Tracking Control Module (TCM) plays a critical role in maintaining precise positioning and motion accuracy by ensuring that the robot adheres closely to the predefined trajectory. Additionally, a Brain Module (BM) consolidates these modules by incorporating intelligent systems such as optimisation algorithms and machine learning techniques, enhancing adaptability and decision-making capabilities within the master controller as depicted in Fig. 8.

Programming plays a vital role in implementing control logic. Students begin by processing sensor data through analog-to-digital conversion, noise filtering, and interrupt handling [21, 26]. Motor control follows, typically involving pulse width modulation (PWM) signals and encoder/potentiometer feedback for closed-loop speed or position regulation [27]. These tasks are often managed using conditional logic, timing loops, or finite state machines (as can be seen in TM module for gait pattern in Fig. 8) to create reactive and sequential robot behaviour [28, 29].

At advanced levels, students explore real-time programming, object-oriented design, and multithreading using professional environments such as STM32CubeIDE, PlatformIO, and Arduino IDE. These tools support code modularity and integration with version control systems [30, 31].

3.2 Controller Architecture and Communication

Modern educational robotics platforms commonly use microcontrollers like Arduino, Raspberry Pi, or ESP32 as central processing units responsible for sensor input, actuator output, and decision-making logic [30, 32]. These embedded systems offer an effective environment for hands-on learning in hardware-software co-design and real-time systems development.

To accommodate increasingly complex robotic applications, hierarchical controller architectures are often adopted. In these configurations, a master controller, typically a Raspberry Pi, oversees high-level planning and communication, while slave controllers, such as Arduino boards, handle dedicated lower-level tasks like motor control or sensor interfacing [33, 34]. This master–slave configuration supports distributed control subsystems and encourages modular design suitable for scalable systems.

Students learn hardware interfacing through GPIO programming and serial protocols such as SPI, I^2C, and UART. Real-time operating systems (RTOS) are introduced to demonstrate task concurrency, priority scheduling, and timing control [30, 32], laying the foundation for scalable embedded robotics. At higher communication layers, wireless Ethernet protocols [35] enable flexible, cable-free data exchange, ideal for remote or dynamic deployments, as shown in Fig. 9. For time-critical applications, fieldbus systems such as Profibus and CAN bus are used. Profibus offers deterministic high-speed communication, while CAN bus provides reliable arbitration and error handling across multiple nodes, supporting fault-tolerant, synchronised control in both educational and industrial settings.

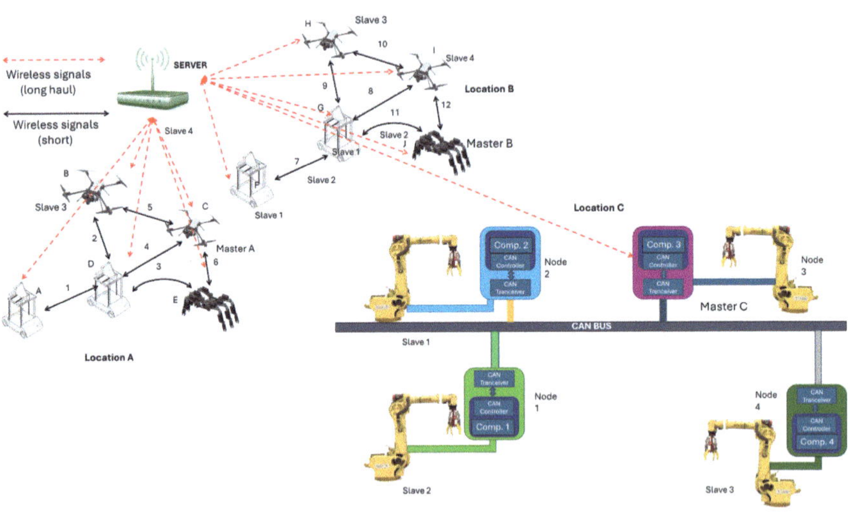

Fig. 9 Example controller networks in robotics

3.3 Sensor Fusion and Path Planning Toward Autonomy

Sensor fusion improves system reliability by combining data from multiple sources. Complementary filters are commonly used to merge accelerometer and gyroscope readings for orientation estimation, whereby a key technique in balance control [21, 24]. For more accurate state estimation, students implement Kalman filters, which account for sensor noise and model prediction.

As students advance, they transition from remote-controlled robots to autonomous systems. Initial projects, such as line-following or maze-solving robots, introduce basic feedback and logic control. Later, they integrate localisation, mapping, and planning algorithms to achieve full autonomy. Techniques like A*, Dijkstra, and RRT are taught for path planning and obstacle avoidance, while simplified SLAM concepts are demonstrated through visual flowcharts [36], helping students understand how robots perceive and navigate unknown environments using probabilistic models. Figure 10 illustrates a sensor fusion approach for autonomous navigation in legged robots. Raw data from sensors such as gyroscopes, LiDAR, radar, and GPS are fused using complementary and Kalman filters to estimate the robot's state. This data is then processed by localisation and mapping modules to construct an environmental model. Path planning algorithms that may be a compilation of A*, RRT, and dynamic window approaches, compute safe and efficient routes in dynamic environments. The resulting information is fed into the Trajectory Module (TM) and Kinematics and Dynamics Module (KDM) to generate precise leg servoing.

3.4 Learning Integration Through Projects

Project-based learning remains a cornerstone of robotics education. Students progress from simple behaviour-based robots to fully autonomous platforms through iterative design and coding. Practical assignments such as maze navigation, object tracking, or balance control reinforce the integration of mechanical design, electronics, and programming [36, 37]. These experiences cultivate system-level thinking and problem-solving skills essential for real-world robotics development.

In summary, control systems and electronics form the operational core of educational robotics. Through hands-on experience, embedded programming, and project-based learning, students gain the foundational knowledge needed to develop intelligent, autonomous robotic systems capable of interacting with real-world environments (Fig. 11).

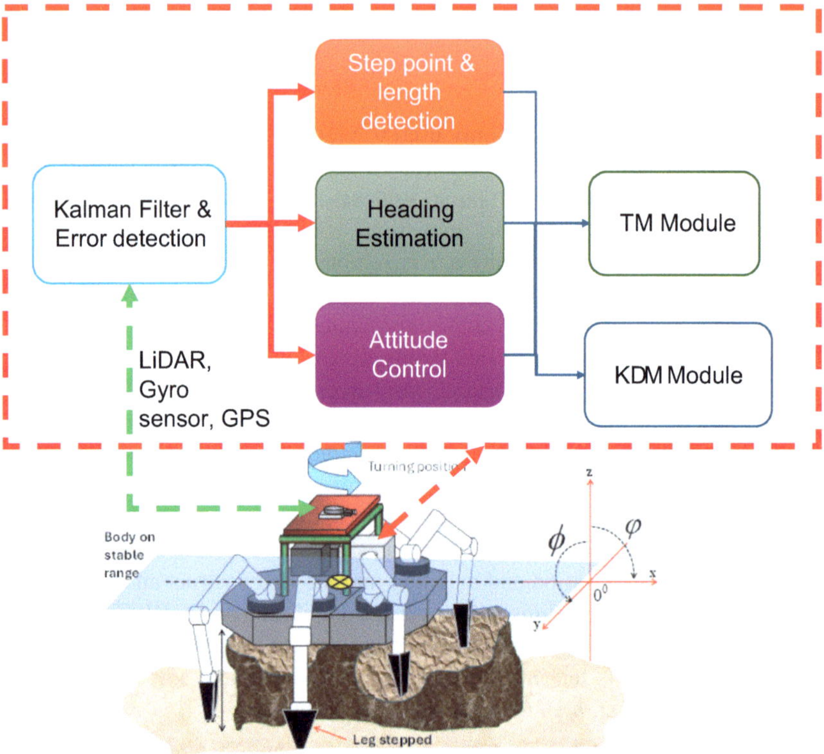

Fig. 10 Integrated sensor fusion and path planning workflow for autonomous robots

4 Conclusion

The integration of mechanical, control, and sensor systems is essential in robotics education, enabling students to design intelligent, functional machines for diverse environments. Understanding robot mechanisms and configurations helps learners choose designs based on task requirements like precision, mobility, and adaptability. Grounded in physics and supported by CAD and simulation tools, mechanical design teaches students to analyse forces, torque, stress, and material behaviour for durable structures. Mastery of control systems and electronics further empowers them to develop responsive, autonomous behaviours using hierarchical architectures, sensor fusion, and programming. Emerging tools like AR, digital twins, and AI-driven assistants enhance visualisation and design efficiency. Through project-based learning, students' progress from basic assemblies to autonomous systems via iterative testing, gaining both theoretical knowledge and hands-on skills crucial for real-world robotics innovation.

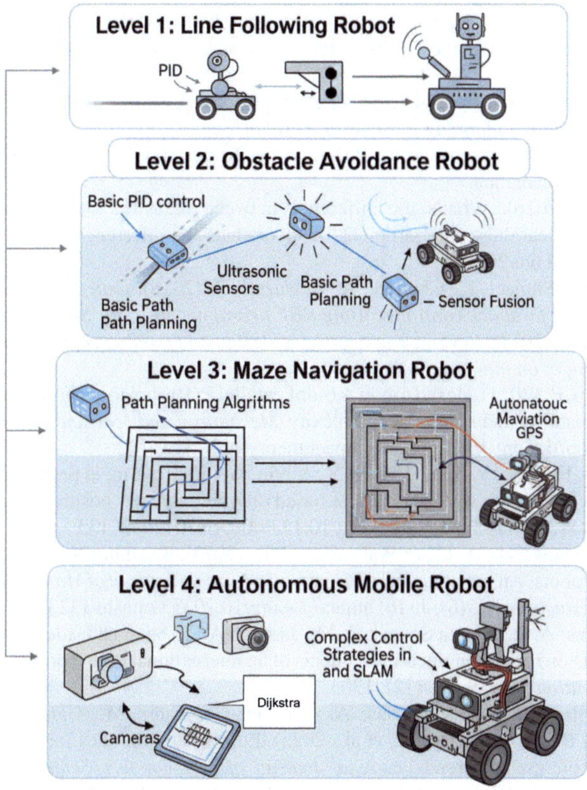

Fig. 11 Example of project-based learning progression in control systems education for mobile robot development

References

1. Varnava, C. (2018). Soft robotics enters uncharted waters. *Nature Electronics, 1*(4), 208–208. https://doi.org/10.1038/s41928-018-0066-4
2. Möller, M., Schappi, A., Buholzer, P., Wegmann, M., Seewer, A., Dubois, F., et al. (2014). Focus project sepios: Riding the wave of progress. ETH Zürich.
3. Arena, P., Pietro, F. D., Noce, A. L., & Patanè, L. (2022). Attitude control in the Mini Cheetah robot via MPC and reward-based feed-forward controller. *IFAC-PapersOnLine, 55*(38), 41–48. https://doi.org/10.1016/j.ifacol.2023.01.131
4. Li, Z., Zeng, J., Chen, S., & Sreenath, K. (2023). Autonomous navigation of underactuated bipedal robots in height-constrained environments. *The International Journal of Robotics Research, 42*(8), 565–585. https://doi.org/10.1177/02783649231187670
5. Wüest, V., Jeger, S., Feroskhan, M., Ajanic, E., Bergonti, F., & Floreano, D. (2024). Agile perching maneuvers in birds and morphing-wing drones. *Nature Communications, 15*(1), 8330. https://doi.org/10.1038/s41467-024-52369-4
6. FESTO Smarbird. (2011). https://www.festo.com/gb/en/e/about-festo/research-and-development/bionic-learning-network/bionic-flying-objects/smartbird-id_33686/2025

7. Yoon, S. P., Jeong, S.-H., Kim, D. K., Yoo, S.-Y., Jun, B.-H., Han, J.-B., et al. (2024). Optimal attitude determination for the CR200 underwater walking robot. *Applied Sciences, 14*(23), 11027.
8. Smith, T., Barreiro, J., Chavez-Clemente, D., Smith, D., & Sunspiral, V. (2008). *ATHLETE's feet: Multi-resolution planning for a hexapod robot.*
9. Peiró, G. O. (2018). Diseño de un Prototipo del Robot Handle de Boston Dynamics con Recurdyn y Mathematica.
10. Gonzalez, C. (2016). What's the difference between industrial robots? https://www.newequipment.com/learning-center/article/22059066/whats-the-difference-between-industrial-robots. Accessed 6/6/2025.
11. Pourrahim, M., Shojaei, K., Chatraei, A., & Nazari, O. (2016). *Experimental evaluation of a saturated output feedback controller using RBF neural networks for SCARA robot IBM 7547.*
12. Tsirogiannis, E. (2017). *Reverse engineering, redesign and topology optimization for additive manufacturing of an industrial robot arm link.*
13. Wu, G., & Niu, B. (2024). Hexad robot: A 6-dof parallel PnP robot to accommodate antagonistic rotational capability and structural complexity. *Mechanism and Machine Theory, 195*, Article 105612. https://doi.org/10.1016/j.mechmachtheory.2024.105612
14. Wang, C., Su, F., Zhao, Y., Liu, H., Guo, Y., & Niu, W. (2020). A rapid performance evaluation approach for lower mobility hybrid robot based on gravity-center position. *Science Progress, 103*(2), 36850420927135. https://doi.org/10.1177/0036850420927135
15. Addido, J., Borowczak, A. C., & Walwema, G. B. (2023). Teaching Newtonian physics with LEGO EV3 robots: An integrated STEM approach. *Eurasia Journal of Mathematics, Science & Technology Education, 19*(6), 1–16. https://doi.org/10.29333/ejmste/13232
16. Navarro Arcas, A. R., Llorca Schenk, J. M., Jara, C. A., & Sentana Gadea, I. (2024). Educational impacts on robotic engineering students of an international online project-based learning course. *Education Sciences, 14*(12), 1305.
17. Martínez Rojas, E., Valencia-Arias, A., Vasquez-Coronado, M., Gamarra, J., Agudelo-Ceballos, E., Benjumea-Arias, M., et al. (2025). Educational robotics for primary education: An analysis of research trends. *Eurasia Journal of Mathematics, Science and Technology Education, 21*, em2602. https://doi.org/10.29333/ejmste/16050
18. Sukackė, V., Guerra, A. O. P. d. C., Ellinger, D., Carlos, V., Petronienė, S., Gaižiūnienė, L., et al. (2022). Towards active evidence-based learning in engineering education: A systematic literature review of PBL, PjBL, and CBL. *Sustainability, 14*(21), 13955.
19. Queiroz-Neto, J. P., Rodrigues, M. A. S., Pereira, M. S., & Gouvea, N. A. (2024). Integrating project-based learning and design thinking: An innovative approach to enhancing hard and soft skills in industrial robotics education. In *2024 IEEE frontiers in education conference (FIE), 13–16 October 2024* (pp. 1–8). https://doi.org/10.1109/FIE61694.2024.10893483
20. Yim, T. S., Leung, C. Y., & Woo, K. T. (2024). Enhancing integrated STEM education through underwater robotics competitions: A project-based learning approach incorporating engineering design processes and Bloom's taxonomy. In *2024 IEEE global engineering education conference (EDUCON), 8–11 May 2024* (pp. 1–5). https://doi.org/10.1109/EDUCON60312.2024.10578695
21. Kumar, N., Arora, U., & Chaudhary, N. (2025). Dynamic equilibrium in robotics: Techniques and applications for self balancing robots. In *2025 IEEE international students' conference on electrical, electronics and computer science (SCEECS), 18–19 January 2025* (pp. 1–6). https://doi.org/10.1109/SCEECS64059.2025.10940101
22. Yang, Z., & Zhang, L. (2024). Balancing control of low-cost wheel-leg robotics for control and robotics education. In *2024 29th International conference on automation and computing (ICAC), 28–30 August 2024* (pp. 1–5). https://doi.org/10.1109/ICAC61394.2024.10718853
23. Hamed, K. A., Kamidi, V. R., Ma, W. L., Leonessa, A., & Ames, A. D. (2020). Hierarchical and safe motion control for cooperative locomotion of robotic guide dogs and humans: A hybrid systems approach. *IEEE Robotics and Automation Letters, 5*(1), 56–63. https://doi.org/10.1109/LRA.2019.2939719

24. Chotikunnan, P., Khotakham, W., Ma'arif, A., Nirapai, A., et al. (2024). Comparative analysis of sensor fusion for angle estimation using kalman and complementary filters. [IMU6050; Kalman Filter; Complementary Filter; Sensor Fusion], *5*(1), 21. https://doi.org/10.31763/ijrcs.v5i1.1674
25. Abulaila, M., Al-Aubidy, K. M., & Alsudi, I. M. (2024). Speed control of permanent magnet synchronous machines: ANFIS design and performance evaluation. In *2024 22nd International conference on research and education in mechatronics (REM), 24–26 September 2024* (pp. 349–353). https://doi.org/10.1109/REM63063.2024.10735555
26. Kaur, D. P., & Kumar, A. (2024). An augmented reality-based user interface design for integrating immersive technology in STEM education. In *2024 2nd International conference on emerging trends in engineering and medical sciences (ICETEMS), 22–23 November 2024* (pp. 361–366). https://doi.org/10.1109/ICETEMS64039.2024.10964981
27. Trehan, M., Gupta, A., & Jung, S. (2019). Implementation of joint control of a 6-DOF manipulator in ROS: KOBOKER for robotics education. In *2019 19th International conference on control, automation and systems (ICCAS), 15–18 October 2019* (pp. 1120–1123). https://doi.org/10.23919/ICCAS47443.2019.8971694
28. Wang, Z. (2025). Application of the PID algorithm in robot. *ITM Web of Conferences, 73,* 01025.
29. Yuan, F., Shen, X., Wu, J., & Yang, W. (2023). Research on simulating application of artificial intelligence in aerospace field. In *2023 2nd International symposium on control engineering and robotics (ISCER), 17–19 February 2023* (pp. 270–273). https://doi.org/10.1109/ISCER58777.2023.00054
30. Beneder, R., Schmitt, P., & Kornyefalvy, C. (2023). A model-based approach for robotics education with emphasis on embedded systems. In *2023 9th International conference on automation, robotics and applications (ICARA), 10–12 February 2023* (pp. 1–5). https://doi.org/10.1109/ICARA56516.2023.10125977
31. Chikurtev, D. S., Chikurteva, A. A., & Spasova, N. R. (2021). Information technologies for development of educational resources in robotics. *IOP Conference Series: Materials Science and Engineering, 1031*(1), Article 012122. https://doi.org/10.1088/1757-899X/1031/1/012122
32. Mustafa, D., Khabour, S. M., & Mustafeh, I. G. (2024). Enhancing real-time embedded system education with self-driving car models. *Human Behavior and Emerging Technologies, 2024*(1), 8578058. https://doi.org/10.1155/2024/8578058
33. Saravanos, A., Li, Y., & Theodorou, E. (2023). *Distributed hierarchical distribution control for very-large-scale clustered multi-agent systems.*
34. Zuo, C., He, K., Shao, J., & Sui, Y. (2024). Self model for embodied intelligence: modeling full-body human musculoskeletal system and locomotion control with hierarchical low-dimensional representation. In *2024 IEEE international conference on robotics and automation (ICRA), 13–17 May 2024* (pp. 13062–13069). https://doi.org/10.1109/ICRA57147.2024.10610081
35. Addie, I., Mohammad Fadhil, A., & Nurulfadzilah, H. (2019). *Robot local network using TQS protocol for land-to-underwater communications.* National Institute of Telecommunications.
36. Sarkisov, Y. S., Coelho, A., Santos, M. G., Kim, M. J., Tsetserukou, D., Ott, C., et al. (2023). Hierarchical whole-body control of the cable-suspended aerial manipulator endowed with winch-based actuation. In *2023 IEEE international conference on robotics and automation (ICRA), 29 May–2 June 2023* (pp. 5366–5372). https://doi.org/10.1109/ICRA48891.2023.10160718
37. Bremgartner, V., Sousa, J. M. d., Natividade, J., Araujo, A., Velozo, H., Costa, A., et al. (2024). WIP: Educational robotics through interdisciplinary project-based learning in the training of teachers and students in professional and technological education. In *2024 IEEE frontiers in education conference (FIE), 13–16 October 2024* (pp. 1–5). https://doi.org/10.1109/FIE61694.2024.10893166

Practical Robotics Projects—Hands-On Learning for Students

1 Why Hands-On Robotics Matters

As the student hunches over a breadboard, flickering his/her eyes between the tangled wire jumpers and the code on the computer screen, the robot that is supposed to follow the black line goes off course. It keeps going in the wrong direction, even though it was programmed to follow the line on the floor. Even though frustrated, the curious student modifies the threshold value of one of the line sensors in the if-else statement and reruns the code. Unfortunately, still no luck. Then, after several tries, with the adjustment of the room lighting and the amendment of the sharp turn line, the robot finally tracks the line. The student is not just fixing the bug, but he/she is learning how to think to solve that problem.

That is the significant difference that robotic hands-on provides compared to traditional learning. Traditional learning is often referred to as lecture-driven or passive learning, in which the teacher is the primary source of the knowledge, guiding the learning process and providing direct instruction during class [1] 2. Furthermore, the curriculum structure is rigid, with a focus on memorisation and repetition of information, which encourages correctness over curiosity [1]. In addition, the students were evaluated through standardised tests and examinations to measure their understanding and retention of the material.

In contrast with robotic topics, active learning in terms of decision-making processes is required. It involves the engagement with the students through hands-on and practical sessions, which significantly enhance their understanding and retention of complex concepts [3]. It requires iteration of efforts, comprehensive experimentation and reflection, transforming failure attempts into feedback and passive learners into active problem-solvers [4]. Therefore, in the robotics projects, the students do not absorb the content solely, but engage in the full cycle of discovery, questioning, building, testing, and improving.

This chapter is initiated based on the idea that hands-on robotic projects are important in bridging the cognitive and pedagogical gap between theory and action.

N. M. Thamrin et al., *Robotics in Education*,
SpringerBriefs in Applied Sciences and Technology,
https://doi.org/10.1007/978-981-95-3548-4_4

It begins by exploring the educational frameworks that support robotics integration, including the curriculum development and experiential learning. Following this, the chapter outlines the key components of a practical curriculum, including the tools, strategies and learning outcomes. Case studies are then presented to demonstrate how such projects are implemented in diverse educational contexts. The discussion continues by analysing how theoretical principles are embodied through practice, the common challenges faced in implementation, and strategies to overcome them. Finally, the chapter concludes by examining the emerging trends in robotic education and how they shape the future of STEM learning.

2 Pedagogical Frameworks Supporting Practical Robotics

The effectiveness of robotics in education relies on the well-established learning theories that emphasise active engagement, iterative problem-solving, and interdisciplinary thinking [5]. One underlying pedagogical structure that sustains meaningful learning is students' excitement while building a functional robotic system. However, to fully harness the educational value of robotics, the curriculum structure must be aligned with frameworks such as project-based learning (PBL) [6, 7], inquiry-based learning [8] and design thinking [9]. These models provide a flexible structure foundation for designing robot activities that teach technical skills through higher-order thinking, collaboration and creativity.

2.1 Project-Based Learning (PBL)

Integrating STEM education with robotics through PBL is highlighted as a method to enhance students' computational thinking, creativity, and problem-solving skills. This approach allows students to apply theoretical knowledge in practical settings, indirectly fostering independent thinking and creativity [10, 11]. PBL is also emphasised as an effective educational strategy that involves students in real projects, promoting engagement and twenty-first-century skills development that passive learning may not achieve. This method is particularly effective in teaching complex concepts such as ROS-based industrial robotics, as it encourages active participation and collaboration among students [12]. Therefore, robotics PBL serves as a platform to cultivate critical thinking, creativity, and collaboration, preparing students for future workforce demands. Despite the benefits, challenges such as appropriate learning activities and comprehensive teacher support are noted. The effectiveness of PBL in robotics education is dependent on well-designed curricula with pedagogical strategies that align with the learning outcomes [11]. Integrating robotics in STEM education enhances technical skills and provides a holistic learning experience that includes emotional, social, and intellectual development [10, 11].

2.2 *Inquiry-Based Learning (IBL)*

IBL has been effectively integrated into robotics education by involving students in hands-on and experiential activities. This approach encourages students to engage deeply with the material, fostering a more interactive and immersive learning environment. Robotics activities incorporating IBL in their curricula have significantly improved student engagement and motivation in STEM fields [13–16]. Robotics education through IBL helps develop the twenty-first-century skills such as problem-solving, critical thinking and teamwork to complete the task [14], 17. Students learn to apply theoretical knowledge to practical problems, bridging the gap between classroom learning and real-world applications. Robotics education through IBL helps develop essential twenty-first-century skills such as problem-solving, critical thinking, and teamwork [14]. Students learn to apply theoretical knowledge to practical problems, bridging the gap between classroom learning and real-world applications. Integrating AI technologies in the IBL framework can enhance the engagement by providing personalised feedback and adapting to individual learning styles [18]. AI-driven analytics and collaborative robots can tailor the learning experience to meet the specific needs and pace of each student, making the learning process more engaging and effective.

2.3 *Design Thinking*

There are five elements in the design thinking framework, namely, (1) empathy and user-centred design, (2) ideation, (3) prototyping and experimentation, (4) collaboration and teamwork, and (5) feedback and reflection. Design thinking starts with understanding the needs and experiences of the users, which, in the context of teaching robotics, means understanding the students' perspectives and learning styles. This principle ensures that the curriculum is tailored to meet the students' needs and interests, making learning more engaging and effective [19, 20]. Then, it promotes brainstorming and the generation of creative ideas. In robotics education, this can be implemented by encouraging students to develop innovative solutions to problems, fostering an environment where creativity is valued and nurtured [21, 22]. After the solution has been selected, the students are encouraged to create prototypes of their robotic designs through an iterative process. It will allow them to test their ideas, learn from failures and improve their development of a robust and reliable robotic system [19, 23]. Robotics projects require knowledge from various fields such as engineering, programming and mechanical design. The interdisciplinary topics help the students collaborate with teammates from various disciplines. By doing this, they will see the connections between different subjects and apply their knowledge comprehensively [24, 25]. Finally, the design thinking process encourages students to reflect on their prototypes by seeking regular constructive feedback from peers and instructors, leading to deeper learning and continuous improvement [21, 26].

3 Essential Components of a Practical Robotics Curriculum

Designing an effective robotics curriculum involves more than selecting the right hardware or programming language; it requires a structure that balances technical complexity with cognitive development. A well-designed curriculum does not just teach students how to assemble components or write code; it supports their growth in critical thinking, collaboration and creative problem-solving. This section outlines a tiered approach to robotics education, starting with foundational logic and mechanics and advancing toward AI-powered autonomous systems. By aligning tools and tasks with specific learning outcomes at each level, educators can deliver robotics learning that is both accessible and intellectually rigorous. Whether using simple microcontrollers or an advanced robotic framework like Robot Operating System (ROS), the main goal is to equip the students with the technical literacy and adaptive mindset needed in the real world.

3.1 Roadmap for Practical Robotics Curriculum

A structured roadmap can be established to develop, sequence, and evaluate robotics projects in educational settings. This roadmap will include tiered project designs, a learning outcomes matrix, and recommended hardware/software tools for each level. Figure 1 shows the roadmap curriculum for teaching robotics.

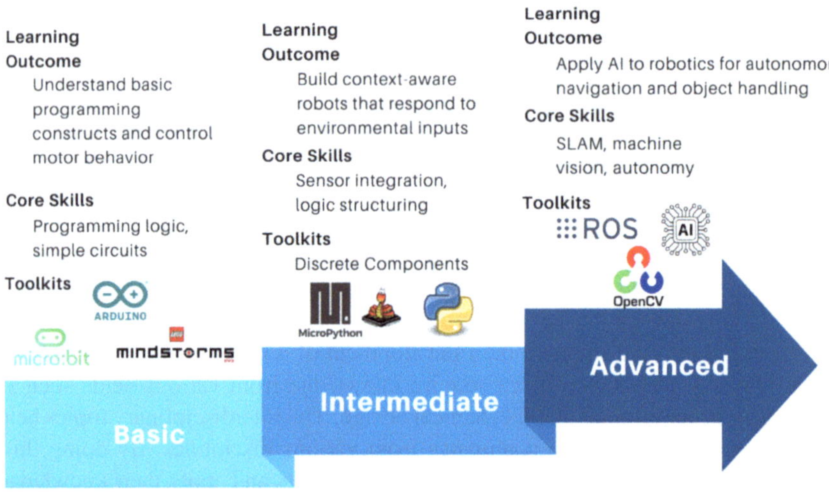

Fig. 1 A tiered progression model for practical robotics education

Figure 1 illustrates a tiered framework for developing a robotics curricula roadmap that aligns with students' cognitive and technical growth. The roadmap is divided into three levels, namely, basic, intermediate, and advanced. This tiered framework maps learning outcomes, core skills, and relevant toolkits at each stage of progression.

3.2 The Basic Level

At the basic level, students are introduced to the core principles of robotics through simple, tangible activities. Common platforms include the BBC Micro: bit, LEGO Mindstorms, and entry-level Arduino kits. Using colourful block programming, these tools are ideal for teaching fundamental programming concepts such as loops, conditionals, and input/output operations. Hands-on activities might involve blinking LEDs, operating simple motors, or building line-following robots using a single sensor. The goal at this stage is to build confidence and spark curiosity. Students understand how software commands affect physical hardware and how logic structures like "if–then" statements drive robot behaviour. Beyond technical skills, learners develop basic troubleshooting strategies and engage in early-stage collaboration through guided group tasks.

3.3 The Intermediate Level

At the intermediate level, students move from single components to working with integrated systems. Platforms such as Arduino, ESP32 and Raspberry Pi with multiple sensors, and simple Python or Micropython programming environments provide the flexibility to design more responsive and adaptable robots. Projects at this stage often involve combining various sensors, such as ultrasonic, infrared, light, and gas sensors, to create a context-aware robot. Students might design robots that avoid obstacles, follow lines with PID control, or respond to environmental changes. This phase introduces basic system architecture thinking, understanding how components interact, how data is collected and processed, and how real-time decisions are made. The learning outcomes extend beyond technical execution. Students are challenged to decompose problems, document their design process, and work more independently in teams. They begin to understand the importance of iterative testing and design optimisation.

3.4 The Advanced Level

In the advanced stage, students are introduced to robotics platforms that support high-level autonomy, vision-based navigation, and AI integration. Tools such as ROS

(Robot Operating System), OpenCV, Jetson Nano, and simulation environments like Gazebo or Webots allow for developing sophisticated robotic behaviours. Typical projects at this level include object recognition using machine learning models, simultaneous localisation and mapping (SLAM), and autonomous decision-making based on sensor data. Students explore sensor fusion, feedback control loops, and real-time data processing concepts. This phase requires learners to apply advanced programming and engineering principles. Students also appreciate robotics' ethical, societal, and design implications, preparing them for real-world innovation and research in industrial automation, environmental monitoring, or assistive robotics.

Therefore, tool selection during the curriculum design is essential as each platform offers different learning options, functionalities and costs. A high-end platform, such as Raspberry Pi has a steeper learning curve and higher cost. However, it provides a powerful computing platform for advanced tasks, such as AI-based implementation, and ROS is the de facto standard for professional-level robotics education and research. Meanwhile, Micro:bit and LEGO kits are ideal for early engagement due to their simplicity and ease of use. On the other hand, Arduino offers more flexibility for customisation and is suitable for intermediate projects. Consequently, it is important to align the tool with the learning objectives, rather than the reverse. The well-chosen platform should facilitate the learning objectives and skills development while promoting deeper exploration, creativity and problem-solving among students.

4 Practical Robotics Projects Implementation (Case Studies)

4.1 Basic Learning

In basic learning, the tools used were from fully integrated kits to modular microcontrollers with simplicity in the programming language. For example, the UERJ Robotics Team employed LEGO EV3 and Arduino by blending the visual programming (colourful block programming) and circuit-based logic to support a wide spectrum of students in public school outreach programs [27]. Similarly, the Edubot project by [28] used the Sparki robot to introduce logical control and sensor-based navigation, as this robot is equipped with built-in sensors and pre-configured mobility features. In contrast, a multiplatform educational robotics course has been introduced using three different programming languages, namely, Scratch, Crumble and Arduino [29]. Various languages are introduced to scaffold students from visual to textual programming in a structured progression. In addition, the Little Engineers project, which focuses on the early childhood learners' experiences (third-grade students) using Ozoblockly, a block-based environment. Due to their young age, the students have shown a strong engagement in the design phase, but struggled with the linear logic, conditionals and loops. This finding is supported by the study of the Sensor Management Team [30], which used Scratch and basic sensors like distance and tilt.

Their study revealed that the beginners were predominantly employed in coventional event-driven code and their innovative idea was limited by the knowledge gap in the type of sensors used.

Regarding project implementation, the pedagogical intent and complexity vary among these studies. UERJ focused on prototyping and demonstration-based activities such as building basic robotic vehicles and automated mechanisms, which strongly align with the school curriculum in physics and math. The project aimed to convert the abstract of the STEM concept into tangible and relatable physical experimentation that helps the students to understand the latter courses better [28]. Meanwhile, Edubot followed a more structured curriculum with 10 instructional sessions that covered a practical obstacle-avoidance robot using an ultrasonic sensor. This project reflected a PBL model by emphasising iterative development and application of theoretical concepts [28]. The multiplatform course incorporated creative tasks using visual programming tools that encouraged students to build interactive systems using sensors and motors through increasingly difficult modules [29]. Findings confirmed that robotic projects have significantly increased their interest and motivation in STEM education through hands-on learning. Furthermore, according to the report, the curiosity towards engineering careers increased, and problem-solving skills improved among the students [27]. The students favoured active learning over the traditional approach, highlighting the motivational power of direct interaction with the technology. Even though they initially struggled with textual syntax in Arduino, their perception of programming improved overall due to the scaffolded nature of block programming.

4.2 *Intermediate Learning*

At the intermediate level, the choice of tools must balance functional complexity with cognitive accessibility. TekTrain, a modular architecture that supports various sensor integrations, such as ultrasonic, camera and acoustic, was developed [31]. It encourages students to self-configure the hardware and components to match the task requirements. Its open-source design and customisable interface make it a more versatile platform for multi-sensor and logic-driven robotics challenges. Similarly, the presented smart motor system by [32] moves beyond beginner programming by inducing a model where sensors are trained, rather than programmed, to trigger the robots' motion. This implementation allows students to learn a feedback-based control system without heavy coding, indirectly fostering conceptual understanding of condition-action systems in robotics. Arduino-based projects remain central at this level due to their expandability. In the National Common Curriculum Base (BNCC) aligned curriculum implementation by de Lima et al. [33], the Arduino platform was used in weekly laboratory sessions to teach students about the real-world applications, such as automated gates and innovative trash bins. These projects require integrating logic, input sensors and actuators in a complete system. In addition, the Learning IoT framework developed by Picanco et al. [34] took a step further by introducing cloud-based code deployment to physical devices, helping students simulate and upload control logic remotely. This is highly relevant in the era of distributed systems and remote laboratories.

Intermediate-level implementation requires significant instructional scaffolding, particularly when students are introduced to unfamiliar technologies or multi-component systems. TekTrain used a spiral development model, incorporating teacher and student feedback to refine each system iteration. This ensured that each learning cycle built meaningfully on the last, aligning well with constructivist principles [31]. Meanwhile, the smart motor initiative emphasised equity and international collaboration, adapting toolkits to fit local constraints. Instructors were not required to teach code syntax directly. Instead, they facilitated a training interface where students "taught" behaviours to the system. This approach broadened participation and empowered less technically confident educators to lead robotics activities successfully [32]. All implementations reported meaningful shifts in student capabilities at the intermediate level. TekTrain students demonstrated increased autonomy and systems-level reasoning, often iterating hardware and software designs without direct prompting. The ability to select appropriate sensors, interpret input data, and adjust control logic was a clear sign of progression beyond basic robotics.

4.3 Advanced Learning

One of the hallmarks of advanced robotics education is the transition to professional-grade tools. In the intelligent robotics master's program at Kazan Federal University, students worked extensively with ROS, Gazebo, URDF, and SolidWorks to model, simulate, and prototype manipulators [35]. The project required learners to develop URDF-based robot descriptions, control logic, and simulation environments for a button-pushing manipulator deployed in elevator systems. Integrating mechanical modelling, kinematics, and real-time systems reflects industry-standard workflows. Similarly, in a Brazilian engineering curriculum, Arduino-based servo control and V-REP simulations were used to develop a planar manipulator robot, guiding students through phases such as CAD design, simulation, prototyping, and control tuning [36]. This hands-on methodology emphasised not just the final product, but the entire engineering pipeline from sketching and modelling to debugging servo behaviour. Projects like MINT-Robo, targeting gifted secondary students, used TurtleBot with ROS, bridging K-12 and undergraduate-level concepts such as localisation, SLAM, and path planning. Despite the learners' younger age, the project pushed them toward algorithmic thinking and modular design strategies often reserved for post-secondary education [37].

Advanced projects are no longer sandboxed exercises. They simulate or directly address real-world systems from industrial manipulators to assistive rehabilitation technologies. The IFAM-LG Industrial Robotics Initiative used a Design Thinking—PBL hybrid approach, involving students in the full lifecycle of a robotic solution deployed within a factory [38]. Students developed solutions for real industrial tasks, reinforcing robotic control, problem scoping, client empathy, and prototype iteration. The Assistive Robotics 5-DoF System project similarly emphasised social relevance. Students built a low-cost upper-limb movement tracker using ESP32 microcontrollers and MPU9250 IMUs, offering practical exposure to sensor fusion, kinematics, and biomedical applications [39]. Meanwhile, in the study of educational robots for specific technical classes, mobile robots were integrated into control and signal processing classes, where students implemented PID controllers and tracked movement using ArUco markers via OpenCV, which is a practical merger of AI and control system theory [40].

Across all reviewed projects, outcomes extended beyond technical skill acquisition. In MINT-Robo, students showed increased confidence in tackling university-level robotics challenges, mastering algorithmic logic and teamwork dynamics [37]. In IFAM's industry-linked projects, evaluations emphasised not only student mastery of industrial robotics tooling, but also their growth in collaboration, communication, and professional decision-making, soft skills increasingly demanded in Industry 4.0 ecosystems. Students in the assistive robotics project gained insights into rehabilitation robotics while also learning how to handle real-world data, synchronise inertial measurements, and apply control logic to socially meaningful tasks.

5 Challenges in Implementation

While robotics education offers immense potential across learning levels, its implementation is fraught with challenges from foundational instruction to advanced system design. At the basic level, resource constraints such as insufficient laptops, limited robotics kits, and a lack of infrastructure often restrict equitable student access and hands-on engagement, particularly in under-resourced public schools [27]. Additionally, students with no prior programming experience face steep initial learning curves, especially when transitioning from visual interfaces to textual languages like C++ or Python, requiring heavy scaffolding and reliance on code examples [29]. Younger learners struggle with computational abstractions such as loops and conditional logic, often defaulting to linear programming approaches without sufficient cognitive support. At the intermediate level, where system integration becomes a focus, challenges shift toward managing multi-component systems, debugging sensor-based logic, and ensuring instructional alignment with increasingly complex tools like Arduino, Raspberry Pi, or IoT frameworks. Students often require structured progression to avoid cognitive overload when learning to fuse hardware control with real-time decision-making [34]. Instructors, particularly those from non-engineering backgrounds, may also feel underprepared to facilitate lessons that involve embedded systems, ROS basics, or cloud-based robotics environments [32]. At the advanced level, projects that involve mobile autonomy, robotic manipulators, or industrial-grade simulations introduce a new set of barriers. The complexity of professional tools such as ROS, Gazebo, or real-time image processing demands not only prior technical knowledge but also significant project planning, hardware calibration, and error handling—elements that can overwhelm even technically inclined students without proper scaffolding [35]. Implementation also faces systemic challenges such as misalignment between university learning goals and industrial timelines, insufficient technical documentation, and limited instructor familiarity with rapidly evolving robotics platforms [37, 39]. Furthermore, maintaining student motivation and resilience is an ongoing concern across levels, especially when early hardware failures, abstract simulations, or debugging hurdles lead to frustration. Collectively, these challenges emphasise the need for scaffolded curriculum design, teacher training, and resource planning tailored to each level of robotics education. Addressing them is essential to unlock robotics' transformative potential in building technical and twenty-first-century skills.

6 Conclusion

Robotics education stands at the intersection of theory and action, offering learners a dynamic path to acquiring technical skills, problem-solving, creativity, and collaboration. Through hands-on projects, students move beyond passive knowledge consumption to actively engage in designing, testing, and iterating real-world systems. This chapter has outlined a structured, tiered approach to robotics education from basic to advanced levels, each aligned with specific tools, learning outcomes, and pedagogical strategies. At the basic level, students build confidence using visual programming and modular kits like Micro:bit and LEGO. Intermediate learners shift toward system integration, working with platforms like Arduino and Raspberry Pi, where they apply logic across sensors and actuators. Advanced students engage in high-autonomy tasks using ROS, OpenCV, and simulation tools, gaining exposure to real-world problem-solving and industry-relevant workflows.

Despite its promise, robotics education is not without challenges. Constraints in resources, instructor readiness, cognitive overload, and technical complexity can hinder implementation. Addressing these barriers requires intentional curriculum design, progressive scaffolding, and professional development for educators. When designed with purpose, robotics projects do more than teach technology. They cultivate adaptive thinkers ready to navigate an increasingly automated and interdisciplinary world. As educational landscapes evolve, integrating robotics meaningfully into curricula offers a powerful way to prepare students for future jobs and lifelong learning in a technology-driven society.

References

1. Sethi, A., Sethi, P. K., & Eren, A. A. (2024). Interdisciplinary learning: Probing into the role of digital technologies in management education. In *Interdisciplinary approaches in management education*. https://doi.org/10.1201/9781003495499-5
2. Yarychev, N. U., & Mentsiev, A. U. (2020). Impact of digital education on traditional education. *Journal of Physics: Conference Series, 1691*(1). https://doi.org/10.1088/1742-6596/1691/1/012132
3. Costa, A. R., Cornelio, E. M. B., Silva, F. R., Batista, F. A., Santos, L. P. C., Oliveira, M. J. A., Koike, C. M. C. C., & Viana, D. M. (2020). Active learning workshops production: Impacts and benefits for engineering students. *International Symposium on Project Approaches in Engineering Education, 10*, 206–213. https://www.scopus.com/inward/record.uri?eid=2-s2.0-85090822315&partnerID=40&md5=c493813422d3458b8b751a168e6497ff
4. Fakaruddin, F. J., Shahali, E. H. M., & Saat, R. M. (2024). Creative thinking patterns in primary school students' hands-on science activities involving robotic as learning tools. *Asia Pacific Education Review, 25*(1), 171–186.
5. Khan, A. (2024). From theory to practice: Implementing interdisciplinary educational strategies in K-12. *Frontiers in Interdisciplinary Educational Methodology, 1*(03), 85–94.
6. Barradas, R., Lencastre, J. A., Soares, S. P., & Valente, A. (2024). Arduino-based mobile robotics for fostering computational thinking development: An empirical study with elementary school students using problem-based learning across Europe. *Robotics, 13*(11), 159.
7. Chen, Y.-K., Chang, C.-C., & Tseng, K.-H. (2015). The instructional design of integrative STEM curriculum: A pilot study in a robotics summer camp. In *Proceedings of 2015 international conference on interactive collaborative learning, ICL 2015* (pp. 871–875). https://doi.org/10.1109/ICL.2015.7318142
8. Bruzgiene, R., Narbutaite, L., Adomkus, T., Pauliute, J., & Kunicina, N. (2022). Inquiry-based learning focused on inclusive education in programming of robots: Challenges, experience, and feedback of first cycle students. In *Inclusive digital education* (pp. 71–95). Springer.
9. Kuo, H.-C., Yang, Y.-T.C., Chen, J.-S., Hou, T.-W., & Ho, M.-T. (2022). The impact of design thinking PBL robot course on college students' learning motivation and creative thinking. *IEEE Transactions on Education, 65*(2), 124–131. https://doi.org/10.1109/TE.2021.3098295
10. Hsu, R. C., & Tsai, T. H. (2022). Assessing the impact of a project-based learning robotics course with integrating of STEM education using content analysis method. *European Journal of STEM Education, 7*(1). https://doi.org/10.20897/ejsteme/12633
11. Lima, B. C., Paiva, H. M., & Maximo, M. R. O. A. (2024). Teaching robotic concepts using project-based learning: A case study. In *EDUNINE 2024—8th IEEE world engineering education conference: Empowering engineering education: Breaking barriers through research and innovation, proceedings*. https://doi.org/10.1109/EDUNINE60625.2024.10500567
12. Hussin, H., Jiea, P. Y., Rosly, R. N. R., & Omar, S. R. (2019). Integrated 21st century science, technology, engineering, mathematics (STEM) education through robotics project-based learning. *Humanities and Social Sciences Reviews, 7*(2), 204–211. https://doi.org/10.18510/hssr.2019.7222
13. Blancas, M., Valero, C., Mura, A., Vouloutsi, V., & Verschure, P. F. M. J. (2020). "CREA": An inquiry-based methodology to teach robotics to children. *Advances in Intelligent Systems and Computing, 1023*, 45–51. https://doi.org/10.1007/978-3-030-26945-6_4
14. Estivill-Castro, V. (2020). Inviting teachers to use educational robotics to foster mathematical problem-solving. *Advances in Intelligent Systems and Computing, 1023*, 248–261. https://doi.org/10.1007/978-3-030-26945-6_22
15. Jou, M., Wu, M.-J., & Wu, D.-W. (2008). Development of online inquiry environments to support project-based learning of robotics. In *Lecture notes in computer science (including subseries lecture notes in artificial intelligence and lecture notes in bioinformatics), 5288 LNAI* (pp. 341–353). https://doi.org/10.1007/978-3-540-87781-3_38

16. Ma, H.-L., Wang, X.-H., Zhao, M., Wang, L., Wang, M.-R., & Li, X.-J. (2020). Impact of robotic instruction with a novel inquiry framework on primary schools students. *International Journal of Engineering Education, 36*(5), 1472–1479. https://www.scopus.com/inward/record.uri?eid=2-s2.0-85095966080&partnerID=40&md5=e0e9e3fd0d61919d7c89234359ca4639
17. Jadhav, S., Iyer, S., & Arya, K. (2024). Enhancing problem-solving in robotics education: Analysis of scaffolds and interactions. In *Proceedings—2024 IEEE international conference on advanced learning technologies, ICALT 2024* (pp. 121–123). https://doi.org/10.1109/ICALT61570.2024.00041
18. Ghofur, A. (2025). From classroom to sustainability: The impact of AI-integrated educational robotics. In *Innovations in educational robotics: Advancing AI for sustainable development.* https://doi.org/10.4018/979-8-3693-6165-8.ch011
19. Díaz-Lauzurica, B., & Moreno-Salinas, D. (2023). Applying design thinking to enhance programming education in vocational and compulsory secondary schools. *Applied Sciences (Switzerland), 13*(23). https://doi.org/10.3390/app132312792
20. Julio, Y. R., Viloria-Núñez, C., Galindo-Jaramillo, J.-F., Mangones, A. P., & Correa, C. C. R. (2024). Innovation strategies for training management with STEAM, IoT, and AI through design thinking. In *ETCM 2024—8th Ecuador technical chapters meeting.* https://doi.org/10.1109/ETCM63562.2024.10746011
21. Arvanitakis, I., Palaigeorgiou, G., & Bratitsis, T. (2022). Evaluating design cards for supporting design thinking in the context of open robotics and IoT competitions. In *Lecture notes in networks and systems, 411 LNNS* (pp. 669–680). https://doi.org/10.1007/978-3-030-96296-8_60
22. Jäggle, G., Lepuschitz, W., Merdan, M., & Vincze, M. (2024). Enhancing sustainability product development and creativity in education: The impact of educational robotics and design thinking on student learning. In *Lecture notes in networks and systems, 901 LNNS* (pp. 391–402). https://doi.org/10.1007/978-3-031-53022-7_39
23. Kolberg, E., Reich, Y., & Levin, I. (2008). Design of design methodology for autonomous robots. In *Lecture notes in computer science (including subseries lecture notes in artificial intelligence and lecture notes in bioinformatics), 5001 LNAI* (pp. 528–539). https://doi.org/10.1007/978-3-540-68847-1_57
24. Chalmers, C., & Nason, R. (2017). Systems thinking approach to robotics curriculum in schools. In *Robotics in STEM education: Redesigning the learning experience.* https://doi.org/10.1007/978-3-319-57786-9_2
25. Plaza, P., Castro, M., Merino, J., Restivo, T., Peixoto, A., Gonzalez, C., Menacho, A., Garcia-Loro, F., Sancristobal, E., Blazquez, M., Diaz, P., Plaza, I., Fondon, I., Sarmiento, A., Civantos, I., Fernandez, C., Lord, S., Rover, D., Chan, R., … Strachan, R. (2020). Educational robotics for all: Gender, diversity, and inclusion in STEAM. In *Proceedings of 2020 IEEE learning with MOOCS, LWMOOCS 2020* (pp. 19–24). https://doi.org/10.1109/LWMOOCS50143.2020.9234372
26. Chahine, I. C., Robinson, N., & Mansion, K. (2020). Using robotics and engineering design inquiries to optimize mathematics learning for middle level teachers: A case study. *Journal on Mathematics Education, 11*(2), 319–332. https://doi.org/10.22342/jme.11.2.11099.319-332
27. Ribeiro Alves, A., de Quadros Albuquerque, T., & Cerqueira Revoredo, T. (2024). UERJ robotics team's educational project. In *2024 Brazilian symposium on robotics and 2024 workshop on robotics in education, SBR-WRE 2024* (pp. 342–347). https://doi.org/10.1109/SBR/WRE63066.2024.10837754
28. Nunes Daltro, I., Silva Santos, J. V., de Sousa, G. B., Martins, R., Pastrana, M. A., & de Souza Baptista, R. (2024). Project Edubot: Exploring pedagogical approaches in high school robotics education. In *2024 Brazilian symposium on robotics and 2024 workshop on robotics in education, SBR-WRE 2024* (pp. 360–364). https://doi.org/10.1109/SBR/WRE63066.2024.10837952
29. Plaza, P., Carro, G., Blazquez, M., Sancristobal, E., Castro, M., & García-Loro, F. (2025). Multiplatform educational robotics course to introduce children in robotics. In *2018 IEEE frontiers in education conference (FIE)* (pp. 1–9).

30. Papigiotis, E., Karvounidis, T., Ladias, A., & Douligeris, C. (2021). Management of sensors in robotic constructions provided by primary school children. In *6th South-East Europe design automation, computer engineering, computer networks and social media conference, SEEDA-CECNSM 2021*. https://doi.org/10.1109/SEEDA-CECNSM53056.2021.9565884
31. Dimitrios, S., Fotios, G., Emmanouil, S., Areti, P., Dimitris, R., & Christos, S. C. (2021). A novel, fully modular educational robotics platform for Internet of Things applications. In *2021 1st Conference on online teaching for mobile education, OT4ME 2021* (pp. 138–145). https://doi.org/10.1109/OT4ME53559.2021.9638892
32. Dahal, M., Kresin, L., Peres, A., Bento Pereira, E., & Rogers, C. (2023). International collaboration to increase access to educational robotics for students. In *Proceedings—Frontiers in education conference, FIE*. https://doi.org/10.1109/FIE58773.2023.10343494
33. de Lima, J. P. R., Filho, E. A. U. M., Costa, F. A. S., & Sarah, S. T. (2024). The integration of education eobotics in high school in the light of the BNCC: Potentialities and challenges—Experience report. In *2024 Brazilian symposium on robotics and 2024 workshop on robotics in education, SBR-WRE 2024* (pp. 279–282). https://doi.org/10.1109/SBR/WRE63066.2024.10837989
34. Picanco, W. D. S., MacHado, G. S., De Carvalho, M. M., Ayres, F. A. D. C., De Medeiros, R. L. P., & De Lucena, V. F. (2022). Learning-IoT: Methodological framework for remote robotics teaching. In *Proceedings—Frontiers in education conference, FIE, 2022-October*. https://doi.org/10.1109/FIE56618.2022.9962451
35. Sulaiman, S., Eryomin, A., Tsoy, T., & Magid, E. (2024). Methodology of designing and modeling projects for robotics and mechatronics course of intelligent robotics master program. In *2024 IEEE 7th Eurasian conference on educational innovation: educational innovations and emerging technologies, ECEI 2024* (pp. 410–413). https://doi.org/10.1109/ECEI60433.2024.10510833
36. Barbosa, A. S., De Oliveira, A. P., Silva, A. R., Boas, A. S. C. A. V., Boas, E. R. V., Felipe De Almeida, F. D. S. A., E Souza, J. V. A., & Boas, M. A. V. (2018). Design and development of a manipulator for educational robotics. In *Proceedings—15th Latin American robotics symposium, 6th Brazilian robotics symposium and 9th workshop on robotics in education, LARS/SBR/WRE 2018* (pp. 522–527). https://doi.org/10.1109/LARS/SBR/WRE.2018.00102
37. Kandlhofer, M., Steinbauer, G., Menzinger, M., Halatschek, R., Kemeny, F., & Landerl, K. (2019). MINT-Robo: Empowering gifted high school students with robotics. In *2019 IEEE frontiers in education conference (FIE)* (pp. 1–5).
38. Queiroz-Neto, J. P., Rodrigues, M. A. S., Pereira, M. S., & Gouvea, N. A. (2024). Integrating project-based learning and design thinking: An innovative approach to enhancing hard and soft skills in industrial robotics education. *Proceedings—Frontiers in Education Conference, FIE*. https://doi.org/10.1109/FIE61694.2024.10893483
39. Queiroz-Neto, J. P., De Sousa, J. M., Pereira, M. S., & Gouvea, N. A. (2023). Advanced training in industrial robotics using project-based learning and design thinking in a partnership between the university and industry. *Proceedings—Frontiers in Education Conference, FIE*. https://doi.org/10.1109/FIE58773.2023.10342896
40. Santos, T. M. B., Amorim, J. S. J. C. C., Mirella, M. M., Campos, V. F., Manhaes, A. G., Lima, J., & Pinto, M. F. (2023). The impact of educational robots as learning tools in specific technical classes in undergraduate education. In *Proceedings—2023 Latin American robotics symposium, 2023 Brazilian symposium on robotics, and 2023 workshop of robotics in education, LARS/SBR/WRE 2023* (pp. 631–636). https://doi.org/10.1109/LARS/SBR/WRE59448.2023.10332959

Robotics and Artificial Intelligence: Autonomous Learning in Education

1 Introduction

In the twenty-first century, education is undergoing a radical transformation, driven by rapid advances in digital technologies. Among these, robotics and artificial intelligence (AI) stand out as particularly influential. Together, they are redefining the way we teach, learn, and interact with knowledge. The rise of intelligent systems capable of adapting to individual learner needs signifies a shift from traditional teacher-centred methods to learner-driven experiences, facilitated by technology [1].

Autonomous learning, a paradigm wherein educational systems possess the capacity to independently deliver, assess, and adjust instruction, has gained traction across educational landscapes. Unlike traditional e-learning platforms that rely on static content delivery, autonomous systems dynamically modify their approach based on real-time analysis of learner input, engagement levels, and performance metrics [2]. These systems are powered by developments in machine learning, natural language processing, and robotics—allowing for adaptive content generation, conversational interfaces, and embodied learning experiences through educational robots [3].

The educational potential of autonomous learning is vast. Students can access personalised instruction at scale, receive timely feedback, and interact with both digital and physical agents that simulate human-like support. Teachers, meanwhile, can leverage data-driven insights to refine pedagogy and allocate more time to human-centric teaching activities. In special education, robotics and AI provide unique pathways for inclusive learning, catering to students with cognitive or behavioural challenges through responsive and emotionally aware systems [4].

Yet, this transition also raises critical questions. To what extent should autonomy be entrusted to machines in shaping a student's educational journey? How do we ensure equity in access to such transformative tools? What ethical frameworks must guide the deployment of AI in learning environments? These questions are central to understanding the true impact of AI and robotics in education.

N. M. Thamrin et al., *Robotics in Education*,
SpringerBriefs in Applied Sciences and Technology,
https://doi.org/10.1007/978-981-95-3548-4_5

This chapter explores the multifaceted implications of robotics and artificial intelligence in enabling autonomous learning. It traces the evolution of these technologies in educational settings, examines practical applications and benefits, and interrogates the challenges, ethical dilemmas, and policy considerations surrounding their use. Ultimately, it aims to contribute to an informed, balanced, and forward-looking perspective on how AI and robotics can serve as enablers of equitable, engaging, and effective education.

2 Evolution of AI and Robotics in Education

Recent studies have underscored this transformation in education driven by the integration of AI and robotics. For instance, Benkhalfallah et al. [2] detail how adaptive technologies and learning analytics have redefined instructional strategies in technologically advanced classrooms. Meanwhile, Shehata et al. [1] emphasise that AI is not merely augmenting traditional practices but fundamentally reshaping how learners engage with content and educators manage learning outcomes.

The evolution of artificial intelligence (AI) and robotics in education is characterised by gradual yet transformative milestones that have reshaped teaching and learning processes over the decades. From the earliest computer-assisted learning environments to modern-day intelligent tutoring systems and humanoid robots, the trajectory of this evolution highlights a growing reliance on data-driven, personalised, and interactive educational experiences [5]. This progression is visually represented in Fig. 1.

The foundational era of educational technology began in the mid-twentieth century with the introduction of programmed instruction, a method based on behaviourist learning theories. This approach used basic machines and later software to deliver sequential learning content, assess responses, and provide immediate

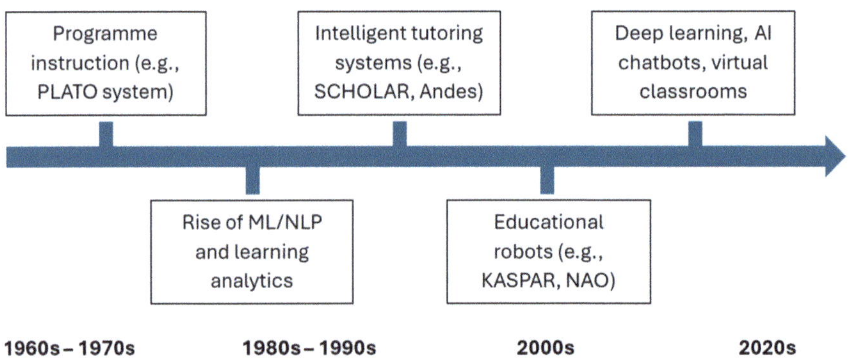

Fig. 1 Timeline showing key phases in the evolution of AI and robotics in education from the 1960s to the 2020s

feedback. One of the earliest examples, the PLATO system developed in the 1960s, enabled students to engage with content at their own pace via terminals linked to a central computer [6].

In the 1980s and 1990s, the rise of personal computers led to the widespread use of computer-assisted instruction (CAI) and the emergence of intelligent tutoring systems (ITS). ITS such as SCHOLAR and the Andes Physics Tutor represented a shift from linear to adaptive instruction, using learner models to provide personalised feedback [7]. These systems improved problem-solving accuracy, reduced time-on-task, and enhanced knowledge retention. Meta-analyses by Ma et al. [8] and Kulik and Fletcher [9] affirmed that well-designed ITS can achieve learning outcomes comparable to those from human tutoring.

The turn of the twenty-first century witnessed a surge in machine learning and natural language processing (NLP) capabilities, which expanded the scope of educational AI. Systems became more sophisticated in recognising learning patterns and predicting future performance. With the advent of big data and cloud computing, platforms could now analyse vast amounts of student interactions and dynamically adapt content. This evolution gave rise to learning analytics, which further supported formative assessments and early intervention strategies [1].

Simultaneously, educational robotics gained momentum. Robots such as KASPAR (developed for children with autism) and NAO (widely used in computer science and language learning) demonstrated how physical agents could foster engagement and support developmental goals. These robots were not only programmable but also equipped with sensors and expressive interfaces, making them suitable for interactive learning tasks [10].

Beyond programmability, these social robots support multimodal learning by combining verbal, visual, and physical cues. Belpaeme et al. [11] reviewed over a decade of studies and found that such robots can enhance learner motivation, attention, and information recall, particularly among young children and students with special needs. Their presence in classrooms has shown promise in scaffolding learning activities, encouraging participation, and modelling socially appropriate behaviour.

Although early adoption focused on primary education and therapeutic contexts, recent years have seen increasing application in STEM disciplines and language acquisition. For instance, robots like NAO and QTrobot have been effectively deployed to teach programming, mathematics, and early literacy. This broadening scope, supported by studies showing improved student motivation and concept retention, underscores the potential of educational robotics to complement human instruction, provided that pedagogical goals are carefully aligned with technological capabilities [11, 12].

In the past decade, the evolution has accelerated with the integration of deep learning, reinforcement learning, and cloud-based AI tools. Systems such as IBM Watson Education, Century Tech, and Squirrel AI offer highly personalised learning experiences, combining real-time performance tracking with adaptive content sequencing. AI-based conversational agents have also entered classrooms, enabling natural dialogue between learners and machines, enhancing critical thinking, and assisting in writing and research tasks [3].

Another noteworthy development is the proliferation of AI-supported MOOCs (Massive Open Online Courses) and virtual classrooms, especially during the COVID-19 pandemic. Originally designed to democratise education by offering free access to high-quality content, MOOCs have evolved significantly through the integration of AI technologies. Platforms such as Coursera, edX, and Khan Academy now leverage AI to personalise course recommendations, predict learner dropouts, assess open-ended responses through NLP, and provide real-time feedback via intelligent tutoring systems. These advancements not only enhance user engagement but also enable instructors to refine content based on learning analytics. The pandemic accelerated the adoption of these tools, establishing AI as a core component of digital pedagogy and reshaping expectations for remote and hybrid learning environments [1].

The integration of AI and robotics into education has not been uniform globally. Countries like South Korea, China, and Singapore have invested heavily in national AI education strategies, embedding intelligent systems into mainstream curricula. Meanwhile, international agencies like UNESCO and the OECD have begun formulating guidelines and policy frameworks to ensure ethical, inclusive, and effective AI deployment in education [13]; OECD, 2021).

Cumulatively, the evolution of AI and robotics in education reflects a shift from static content delivery to dynamic, responsive, and context-aware learning ecosystems. These technologies have matured from supporting tools into autonomous agents capable of facilitating meaningful learning experiences across cognitive, emotional, and social dimensions. The next sections of this chapter will explore how these developments are being applied practically in classrooms and their implications for educators and learners alike.

3 Autonomous Learning Systems

Autonomous learning systems refer to AI-driven educational platforms and agents capable of independently adjusting their instructional strategies in response to learner behaviours and performance metrics. Unlike static digital tools, these systems continuously adapt by leveraging real-time data and past learner interactions to offer personalised, context-aware educational experiences [14].

What sets autonomous learning apart from basic adaptive e-learning is its reliance on artificial intelligence to make real-time pedagogical decisions with minimal reliance on pre-programmed logic. These systems demonstrate levels of autonomy in diagnosing learner needs, selecting content, sequencing instructional tasks, and refining delivery based on engagement or performance metrics [5]. As such, they mark a significant evolution toward data-responsive and learner-centred environments.

At their core, autonomous learning systems integrate a blend of technologies—namely machine learning algorithms, reinforcement learning protocols, natural language processing (NLP), and sensor-based robotics. Machine learning enables these systems to identify learning patterns and predict future learner needs, while reinforcement learning allows agents to optimise feedback mechanisms through trial-and-error. NLP supports natural dialogue interfaces for text- or voice-based interactions, and embedded sensors in robotic platforms allow them to respond to gestures, expressions, and environmental cues [2, 3]. These core components are visualised in Fig. 2, which outlines the input-processing-output cycle of a typical autonomous learning system.

These capabilities have led to a paradigm shift from instructor-led models toward more student-centric learning. For example, platforms like Squirrel AI in China use adaptive learning engines to diagnose misconceptions and tailor learning paths accordingly. Similarly, intelligent virtual tutors embedded within platforms such as

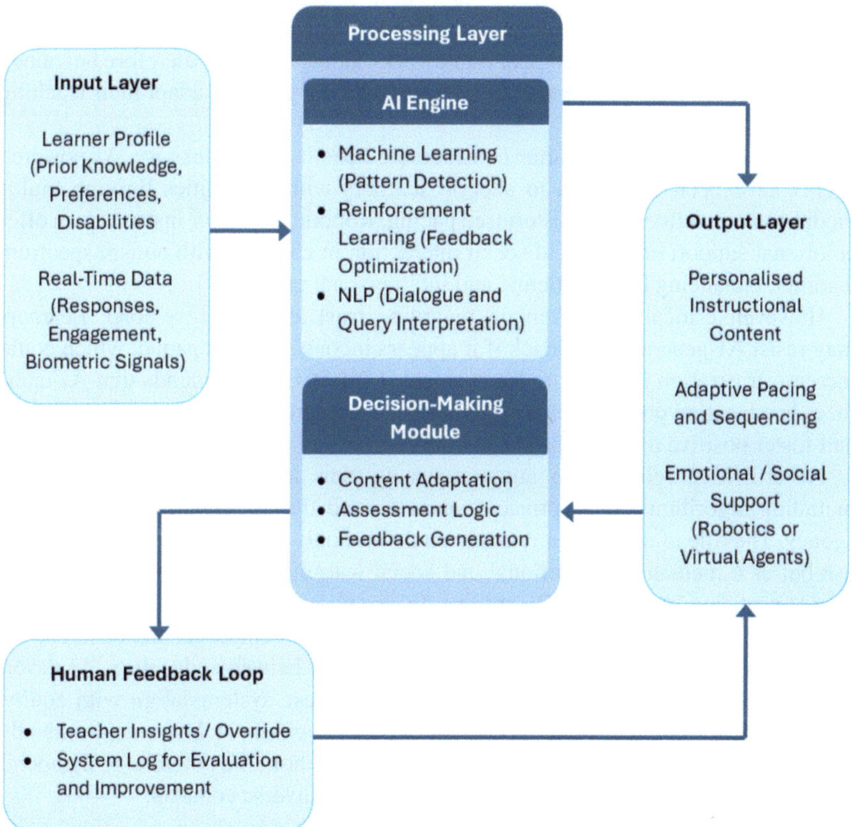

Fig. 2 Conceptual diagram showing the input-processing-output cycle of an AI-driven autonomous learning system

Carnegie Learning and IBM Watson Education provide just-in-time interventions based on real-time assessments, enabling learners to progress at their own pace [5].

In a practical classroom example, a secondary science class in Singapore employs an AI-powered assistant that adjusts the complexity of physics problems in real-time while prompting students with guided questions to promote critical thinking. Biometric feedback—such as gaze tracking and posture—can further personalise pacing, offer breaks or collaborative tasks when needed. Such implementations illustrate the range of contextual responsiveness these systems can achieve.

The outcomes associated with autonomous learning systems are promising. Research by Holmes et al. [6] and Dong et al. [15] indicates that adaptive AI tools can enhance motivation, improve knowledge retention, and increase academic achievement, especially among underperforming students. These tools are particularly effective in delivering formative assessments and personalised feedback, both of which are critical for self-regulated learning.

Importantly, the implementation of autonomous learning systems must be accompanied by sound instructional design and continuous evaluation. As pointed out by the OECD [16], technology alone does not improve learning outcomes unless integrated within coherent pedagogical frameworks. Educators must therefore be trained not only to use AI tools, but also to interpret data insights and adapt their teaching strategies accordingly.

Moreover, such systems offer considerable benefits for inclusivity. AI-powered agents have been developed to support learners with disabilities through multimodal content delivery and customised pacing. Robotic tutors, for instance, can offer emotional support and scaffold social interaction in children with autism spectrum disorder, enhancing both academic and interpersonal growth [4].

However, critical issues remain regarding trust and cognitive load. Learners may resist AI-generated feedback if it appears inconsistent or opaque, which could increase frustration and reduce engagement. UNESCO recommends that AI interfaces in education prioritise explainability and transparency to maintain learner trust and foster positive interactions [13].

Nevertheless, reliance on autonomous systems also presents broader risks, including algorithmic bias, privacy concerns, and the marginalisation of teacher agency. The role of the teacher remains vital—not only as a facilitator of technology use but as a mediator of emotional and social learning experiences that machines cannot replicate. As UNESCO cautions, the goal is not to replace teachers, but to augment their roles through ethically designed, learner-centred technology [13].

Future development will depend on collaboration between educators, AI developers, ethicists, and policymakers to ensure that these systems align with equity, safety, and pedagogical integrity. When deployed within inclusive and ethically grounded frameworks, autonomous learning systems hold the potential to democratise education and elevate learning outcomes across diverse contexts.

4 Applications in the Classroom

The classroom serves as the critical intersection where the theoretical aspirations of autonomous learning systems are tested against the dynamic realities of teaching and learning. In diverse educational settings—from early childhood education to higher learning—autonomous AI-driven tools and robots are now being integrated to enhance engagement, differentiate instruction, and support formative assessment. Their application is not monolithic; rather, it is shaped by contextual needs, available infrastructure, and curriculum design.

One of the most prominent uses of autonomous systems in classrooms is adaptive content delivery. AI-enabled platforms such as Century Tech and Knewton use learning analytics to continuously evaluate student input and automatically modify lesson pathways. These systems personalise learning in real-time, ensuring that struggling learners receive remediation while advanced students can accelerate without waiting for the rest of the class. This dynamic pacing promotes a mastery-based learning model, which is increasingly viewed as a remedy to one-size-fits-all instruction [14, 16].

Autonomous systems also support collaborative and project-based learning. AI-powered classroom assistants can group students based on complementary skills or learning styles, facilitate peer feedback, and recommend roles in team-based tasks. In STEAM (Science, Technology, Engineering, Arts, and Mathematics) subjects, AI is often paired with robotics kits to enable exploratory, hands-on learning. For example, students might use programmable robots to simulate ecological systems or model planetary motion—applying theory through creative experimentation [12].

Moreover, intelligent tutoring systems (ITS) offer targeted, one-on-one support within classrooms. These systems serve as digital tutors, responding to learners' misconceptions, offering scaffolded hints, and tracking cognitive progress. Carnegie Learning's MATHia, for instance, uses a cognitive model to guide students through algebraic reasoning, adapting to their problem-solving approaches and logging progress for the teacher's dashboard [6, 8].

In addition to supporting core subjects, autonomous systems are playing an increasing role in promoting inclusion and accessibility. This diversity of applications is summarised in Table 1. For language instruction, natural language processing capabilities in AI platforms like ELSA Speak and Duolingo enhance oral fluency and pronunciation through immediate feedback. These tools provide a safe environment for repetition and correction, which is particularly beneficial for shy or multilingual learners. In multicultural classrooms, AI chatbots can also serve as real-time translators, breaking down language barriers and fostering inclusivity [13].

These tools provide a safe environment for repetition and correction, which is particularly beneficial for shy or multilingual learners. In multicultural classrooms, AI chatbots can also serve as real-time translators, breaking down language barriers and fostering inclusivity [13].

Crucially, the teacher remains the orchestrator of all these technologies. In classrooms where autonomous systems are used effectively, educators act as interpreters

Table 1 Classroom applications of autonomous learning systems

Application type	Tool/System	Educational benefit
Adaptive content delivery	Century Tech, Knewton	Personalised pacing, mastery learning
Intelligent tutoring	MATHia, Carnegie learning	Scaffolding, real-time feedback
Language learning	ELSA Speak, Duolingo	Pronunciation, fluency, multilingual support
Collaborative learning	AI grouping assistants	Peer feedback, role assignment
Inclusive access	NAO, Pepper, AI chatbots	SEN support, socio-emotional scaffolding
Teacher integration	Dashboards, Analytics tools	Intervention planning, differentiation

of AI feedback, adjusting instruction based on analytics dashboards, refining group configurations, or planning interventions. Professional development programmes must therefore accompany implementation, enabling teachers to align AI tools with pedagogical goals rather than rely on them passively [16].

Despite these advancements, practical challenges remain. Infrastructure gaps, unequal access to devices, and ethical concerns around surveillance and data privacy must be addressed for equitable implementation. Schools in lower-resourced settings may require customised, low-bandwidth solutions. Moreover, schools must establish clear guidelines on AI's pedagogical role, ensuring transparency, explainability, and fairness in autonomous decision-making [13].

Broadly speaking, when guided by pedagogical intent and supported by inclusive infrastructure, classroom applications of autonomous learning systems can serve as transformative tools for personalised, equitable, and future-ready education. Their full potential, however, hinges on continuous professional development, ethical design, and thoughtful alignment with local educational needs. As educators and institutions become more adept at integrating these systems, they stand to not only improve academic outcomes but also redefine the role of technology in human development.

5 Benefits of Autonomous Learning in Education

The integration of autonomous learning systems into educational practice brings a wide array of benefits that extend beyond mere efficiency. These systems provide not only scalable solutions for diverse learning needs but also new pathways for equity, engagement, and empowerment in education. Grounded in machine intelligence and driven by real-time data, autonomous learning systems redefine how students interact with content, instructors, and their own learning processes.

One of the most significant advantages of autonomous learning systems is their ability to deliver personalised instruction at scale. Traditional classroom models often struggle to accommodate diverse learner profiles, particularly in mixed-ability settings. Autonomous systems respond dynamically to each student's pace, preferences, and performance history—customising content delivery to optimise engagement and comprehension. According to Holmes et al. [6], adaptive learning systems significantly enhance learner outcomes by delivering tailored feedback and differentiated support, especially for students who may otherwise fall behind in standardised instruction.

Another major benefit is the facilitation of self-regulated learning. By receiving timely feedback and performance analytics, students develop greater metacognitive awareness and take active ownership of their educational journey. These systems enable learners to track progress, set goals, and self-correct based on AI-driven suggestions—skills closely aligned with twenty-first-century learning competencies [16]. Moreover, such systems promote resilience by allowing learners to revisit content multiple times without fear of judgement or failure.

Autonomous systems also enhance formative assessment and instructional plan ning. Teachers can use analytics dashboards to interpret real-time data from student interactions, identifying learning gaps and misconceptions. This supports more targeted interventions, timely feedback, and informed instructional decisions. In this way, autonomous learning tools extend—not replace—the teacher's role by amplifying their ability to personalise learning for each student [5].

In terms of accessibility, autonomous learning systems reduce barriers for learners with disabilities or those in remote locations. Text-to-speech, gesture recognition, and adaptive interfaces are just a few examples of how AI-powered platforms provide inclusive educational experiences. These tools are particularly effective in supporting neurodiverse learners, students with mobility impairments, or those who require alternative forms of communication [4].

Emotional and social learning are also positively impacted. With the aid of empathetic robots or AI avatars, learners can practice emotional regulation, receive positive reinforcement, and develop social cues in a controlled environment. For students with social anxiety or developmental delays, these platforms can offer a safe space to rehearse communication and build confidence before interacting in larger groups [11]. The key advantages of such systems across various learning dimensions are summarised in Table 2, receive positive reinforcement, and develop social cues in a controlled environment. For students with social anxiety or developmental delays, these platforms can offer a safe space to rehearse communication and build confidence before interacting in larger groups [11].

From a systemic perspective, autonomous learning tools increase instructional equity by offering high-quality educational resources across demographic and geographic boundaries. AI-enabled MOOCs and cloud-based platforms bring personalised learning to under-resourced schools, rural areas, and conflict-affected zones. According to UNESCO, such platforms are instrumental in reducing global disparities in access to quality education [13].

Table 2 Key benefits of autonomous learning in education

Benefit area	Description	Supporting feature/Example
Personalised instruction	Adapts to learner pace, performance, and preferences	Squirrel AI, IBM Watson Education
Self-regulated learning	Promotes goal-setting, metacognition, and learner autonomy	Real-time analytics, adaptive feedback
Formative assessment	Provides data-driven insights for teacher intervention	Dashboards, performance tracking
Accessibility and inclusion	Supports learners with disabilities or in remote areas	Text-to-speech, gesture interfaces
Emotional/social support	Builds confidence, empathy, and interaction skills in safe settings	Robots like NAO, AI avatars
Instructional equity	Delivers quality learning across geographic and socioeconomic barriers	MOOCs, AI-enhanced cloud platforms
Curriculum innovation	Enables modular and competency-based learning linked to learner profiles	Performance mapping, AI curriculum tools

Furthermore, these systems can foster innovation in curriculum design. Educators are now empowered to implement modular, competency-based, and interdisciplinary learning structures. Through continuous learner profiling and performance mapping, curricula can evolve to suit both individual aspirations and societal needs, aligning better with workforce expectations and lifelong learning trajectories [15].

Despite these strengths, it is important to acknowledge that the effectiveness of autonomous systems depends on thoughtful implementation. Without appropriate pedagogical frameworks, teacher training, and ethical safeguards, the benefits of AI may not be fully realised. Issues such as algorithmic bias, over-reliance on automation, and inadequate technological infrastructure must be proactively addressed. Nonetheless, when well-integrated, autonomous learning systems represent one of the most promising educational innovations of the twenty-first century.

Taken together, autonomous learning technologies bring transformative benefits by enhancing personalisation, improving learning equity, supporting inclusive practices, and enabling data-informed teaching. These systems not only meet learners where they are but help propel them towards higher levels of engagement, achievement, and lifelong adaptability.

6 Challenges and Limitations

While autonomous learning systems offer promising benefits, their integration into education is not without challenges. These limitations are technical, pedagogical, ethical, and socioeconomic in nature. Understanding and addressing these concerns is critical to ensuring that such technologies serve as tools for equity and enhancement rather than exclusion or disruption.

One of the foremost challenges is algorithmic bias. Autonomous systems rely on machine learning algorithms trained on datasets that may reflect historical and cultural inequalities. If unchecked, these systems can reinforce stereotypes or marginalise learners who fall outside dominant behavioural or linguistic norms. For instance, language processing algorithms may struggle with non-standard dialects or regional expressions, leading to inaccurate assessments or disengagement. As UNESCO cautions, developers and educators must prioritise fairness and transparency by testing systems across diverse learner populations [13].

Another limitation lies in data privacy and security. Autonomous systems continuously collect and analyse large volumes of student data to inform personalised recommendations. Without robust data governance frameworks, such practices pose significant risks. Breaches in data confidentiality or misuse of learning analytics can compromise student trust and institutional credibility. This is particularly concerning when data is shared with third-party vendors or stored on unsecured cloud platforms. Institutions must develop clear policies on data collection, retention, and ethical use to mitigate these risks [16].

From a pedagogical perspective, over-reliance on automation may diminish the role of critical thinking and human interaction in education. While AI can provide efficient content delivery and performance feedback, it cannot fully replicate the nuances of human judgement, empathy, and contextual awareness. There is a danger that autonomous systems could shift the focus toward algorithmically optimised outcomes at the expense of holistic development. Teachers must remain central as facilitators, interpreters, and ethical stewards of these technologies.

Another notable barrier is infrastructural inequality. Autonomous learning technologies often require reliable internet access, advanced hardware, and regular maintenance—resources that are not equally available across all regions or institutions. This digital divide can widen existing educational disparities, particularly in low-income or rural settings. UNESCO highlights the need for inclusive design and scalable, low-bandwidth alternatives to ensure broad accessibility [13].

The complexity of system implementation also presents a challenge. Successful integration of autonomous learning tools requires not only the technology itself but also a supportive institutional ecosystem. This includes teacher training, curriculum adaptation, IT support, and ongoing monitoring. Without adequate preparation and stakeholder involvement, even well-designed systems may fail to gain traction or demonstrate impact [5].

Ethical concerns surrounding autonomy in decision-making must also be addressed. When AI systems make instructional choices or behavioural recommendations, questions arise about accountability and explainability. Who is responsible if a student is misdiagnosed by an AI tutor or receives biased feedback? Transparent system design, human oversight, and recourse mechanisms are essential to maintaining trust and accountability in educational environments [13].

Finally, there is a cultural dimension to the adoption of autonomous learning systems. Attitudes toward AI in education vary across regions, influenced by values, traditions, and historical relationships with technology. In some contexts, educators may resist automation due to fears of professional displacement or reduced pedagogical autonomy. Building trust through participatory design, localised training, and context-aware implementation strategies can foster acceptance and ownership.

Looking ahead, while autonomous learning systems hold transformative potential, their deployment must be informed by a critical understanding of their limitations. Addressing these challenges requires interdisciplinary collaboration, inclusive policymaking, and a firm commitment to ethical, equitable, and learner-centred educational practices.

7 Ethical and Policy Considerations

The deployment of autonomous learning systems in education necessitates careful ethical and policy deliberation. These systems, while offering transformative benefits, also raise critical questions about fairness, accountability, transparency, and inclusivity. For educational institutions and policymakers, addressing these concerns is essential to ensure that AI technologies align with the broader goals of education equity and human-centred development.

A primary ethical consideration is the transparency and explainability of AI decision-making. Autonomous systems often operate using complex algorithms that are not easily interpretable by educators or students. This opacity can undermine trust and hinder users from understanding how learning paths or feedback are determined. According to UNESCO, explainable AI is crucial for upholding educational integrity, particularly when systems influence high-stakes decisions such as academic placement or personalised interventions [13].

Data ethics also play a central role. Autonomous learning platforms collect extensive behavioural and performance data to tailor content. While this can enhance learning outcomes, it also introduces risks of surveillance, consent violations, and data misuse. Institutions must adopt rigorous data protection policies that include informed consent, anonymisation protocols, and clear data-sharing practices. Aligning with global frameworks such as the General Data Protection Regulation (GDPR) can guide ethical data handling in educational contexts [16].

Equity in access remains both an ethical and policy concern. Without proactive inclusion strategies, AI systems may exacerbate existing disparities. Students from marginalised communities, or those with disabilities, may face barriers to engagement due to language limitations, interface design, or connectivity issues. Policy frameworks must therefore mandate accessibility standards and promote the development of culturally and linguistically responsive AI tools [13].

Another significant concern is algorithmic accountability. In cases where AI-generated feedback or recommendations lead to negative educational outcomes, institutions must determine who bears responsibility—the developer, the educator, or the system provider. Establishing robust governance structures, including AI ethics boards and audit mechanisms, is vital for ensuring that automated decisions are monitored and contestable.

Teacher agency must also be protected. As AI systems take on greater instructional roles, there is a risk of reducing educators to passive monitors. Ethical implementation requires that teachers retain control over pedagogical choices, interpret AI outputs meaningfully, and adapt technologies to fit student needs. Professional development and participatory design approaches can ensure that AI enhances, rather than overrides, educator expertise [14]. These considerations, along with other major ethical and governance priorities, are summarised in Table 3.

Finally, policy alignment is necessary to support sustainable and ethical AI integration. National and institutional policies must address procurement standards, interoperability, long-term funding, and curriculum integration. International bodies like UNESCO and the OECD have begun developing guidelines, but local adaptation is crucial. Context-specific strategies that consider infrastructure, cultural values, and educational priorities will be key to meaningful implementation.

Table 3 Ethical and policy considerations for autonomous learning systems

Ethical/Policy concern	Description	Recommended response
Transparency and explainability	Opaque decision-making can reduce trust and hinder understanding	Develop explainable AI systems and user-friendly dashboards
Data ethics	Risk of surveillance, consent issues, and misuse of personal data	Implement GDPR-aligned data protection policies
Equity in access	Marginalised learners may face interface, language, or connectivity barriers	Mandate inclusive design and linguistic responsiveness
Algorithmic accountability	Unclear responsibility for AI-generated decisions or outcomes	Establish ethics boards and AI audit mechanisms
Teacher agency	Risk of diminishing educator control and professional judgement	Retain pedagogical authority and promote co-design
Policy alignment	Gaps in procurement, funding, and curriculum integration	Create adaptive, context-aware education policies

Ultimately, embedding ethics and policy considerations into the design and deployment of autonomous learning systems is not a peripheral task—it is a foundational imperative. Only through transparent, inclusive, and accountable practices can the potential of AI in education be realised in a way that honours the dignity and rights of all learners.

8 Case Studies

To understand the practical implications of autonomous learning systems, it is useful to examine how different educational institutions and regions have adopted and adapted these technologies. The following case studies illustrate varied approaches to integrating AI and robotics into classroom settings, policy frameworks, and learner support systems. These examples also shed light on contextual factors that influence success and highlight replicable strategies for scaling AI-driven innovation.

8.1 Case Study 1: Squirrel AI—Adaptive Learning in China

Squirrel AI is one of China's most advanced adaptive learning platforms, leveraging AI to diagnose learning gaps and customise instructional content. The system is widely used in after-school tutoring centres, where it analyses student responses to tailor pacing, provide targeted practice, and dynamically modify learning pathways. According to Wang et al. [17], Squirrel AI has shown significant improvements in mathematics achievement among students from underperforming regions. Its model combines reinforcement learning and learning analytics to offer near real-time feedback and predictive diagnostics, creating a high degree of personalisation that rivals one-on-one tutoring.

8.2 Case Study 2: Century Tech—AI-Powered Personalisation in the UK

Century Tech is an AI-based platform implemented in hundreds of schools across the UK. It uses machine learning to recommend lessons, identify misconceptions, and provide micro-assessments. Teachers benefit from data dashboards that help identify students who need intervention. A recent evaluation by the Education Endowment Foundation [18] found that schools using Century Tech reported improved learner engagement, reduced marking load for teachers, and greater confidence in delivering differentiated instruction. The platform has also been lauded for aligning with the

UK's digital education strategy and supporting remote learning continuity during the COVID-19 pandemic.

8.3 Case Study 3: Pepper Robot in Japan's Special Needs Education

Japan has integrated humanoid robots such as Pepper into classrooms for children with special educational needs. Developed by SoftBank Robotics, Pepper is programmed to recognise emotions, facilitate routines, and model social behaviours. According to a study by Kumazaki et al. [19], students on the autism spectrum demonstrated increased attentiveness and participation when working alongside Pepper. Teachers also reported improvements in peer interaction and a decline in disruptive behaviours. These outcomes suggest that socially assistive robots can serve as effective mediators in inclusive learning environments.

8.4 Cross-Case Observations

Despite variations in geography and scope, these case studies reveal several commonalities. First, successful implementations are grounded in strong pedagogical design and teacher training. Second, real-time analytics and adaptive content delivery are central features across all platforms. Third, policy support—whether national (as in the UK and Japan) or institutional (as in local education authorities)—plays a pivotal role in sustaining innovation.

However, challenges also persist. These include managing infrastructure demands, addressing data privacy concerns, and avoiding over-reliance on automated decisions. Culturally responsive design and stakeholder engagement remain essential to ensure that AI systems align with learner needs and contextual realities.

As these case studies illustrate, autonomous learning technologies are not uniform solutions but adaptable systems that require intentional planning and continuous evaluation. Their impact depends not only on technological sophistication but also on human insight, ethical foresight, and systemic coherence.

9 Future Trends and Recommendations

As autonomous learning technologies continue to evolve, several emerging trends are poised to shape the future of education. These trends are grounded in ongoing advancements in AI capabilities, growing emphasis on equity, and the need for lifelong learning in a rapidly changing world. Understanding and responding to these trends will be critical for educators, policymakers, and technologists aiming to design inclusive and future-ready educational ecosystems.

One major trend is the integration of generative AI into autonomous learning platforms. Tools such as ChatGPT and other large language models (LLMs) are being explored to facilitate more natural dialogue, generate context-specific content, and offer sophisticated feedback. This shift towards conversational and generative AI suggests a future where autonomous systems serve not only as tutors but also as co-creators of knowledge, offering personalised guidance and fostering inquiry-based learning.

Another emerging direction is the development of emotionally intelligent AI. As research into affective computing progresses, future systems are expected to be capable of detecting and responding to learners' emotional states. This could significantly enhance engagement, especially for students who struggle with motivation or social-emotional challenges. Emotion-aware systems may also play a vital role in inclusive education, offering responsive support to neurodiverse learners or those with behavioural needs.

Furthermore, there is a growing emphasis on interoperable and modular AI systems that can be customised across different educational contexts. Instead of monolithic platforms, future systems may allow educators to combine best-in-class AI tools tailored to specific pedagogical goals. This modularity will support scalability, flexibility, and more targeted innovation aligned with local curricula and teaching practices.

In response to these trends, several recommendations emerge. First, educators should be involved in the co-design and continuous evaluation of AI tools to ensure pedagogical alignment and usability. Second, governments and educational institutions should invest in digital infrastructure and professional development to bridge the readiness gap. Third, international collaboration is needed to establish standards for ethical AI in education, including fairness, transparency, and accountability.

As we look to the future, the role of autonomous learning systems will extend beyond content delivery towards supporting holistic, adaptive, and equitable learning experiences. By anchoring these technologies in human-centred design and inclusive policy, we can unlock their full potential in shaping the future of education.

10 Conclusion

The incorporation of robotics and artificial intelligence into education represents a fundamental shift in how teaching and learning are conceptualised and delivered. These technologies offer more than automation as they enable personalised learning pathways, support inclusive education, and empower both students and educators through real-time feedback and intelligent guidance. From adaptive platforms to socially assistive robots, autonomous learning systems are steadily reshaping the educational landscape.

This chapter has explored the promise of autonomous systems to enhance engagement, differentiate instruction, and foster equitable access to quality learning. Case studies from China, the UK, and Japan demonstrate that successful implementation depends on more than just technological capability, as it requires pedagogical alignment, ethical foresight, and policy support. At the same time, challenges such as algorithmic bias, infrastructure disparities, and the need to preserve teacher autonomy underscore the importance of thoughtful integration.

Moving forward, the focus must remain on ensuring that these systems are designed and deployed in ways that uphold human dignity and educational integrity. Teachers must remain central to the process, supported rather than supplanted by AI. Through inclusive design, robust governance, and collaborative innovation, autonomous learning technologies can help create adaptive, learner-centred environments that prepare students for the demands of a dynamic, interconnected world.

References

1. Shehata, B., Tlili, A., Huang, R., Adarkwah, M. A., Liu, M., & Chang, T. (2024). How are we doing with student-centered learning facilitated by educational technologies? A systematic review of literature reviews. *Education and Information Technologies, 29*, 7813–7854. https://doi.org/10.1007/s10639-023-12112-w
2. Benkhalfallah, F., Laouar, M. R., & Benkhalfallah, M. S. (2024). Examining adaptive e-learning approaches to enhance learning and individual experiences. *Acta Informatica Pragensia, 13*(2), 327–339.
3. Kaouni, M., Lakrami, F., & Labouidya, O. (2024). The use of artificial intelligence and natural language processing in e-learning platforms: An exploration of potentials and limitations. *International Journal of Information Science and Technology, 8*(4), 1–9. https://doi.org/10.57675/IMIST.PRSM/ijist-v8i4.258
4. Papakostas, G. A., Sidiropoulos, G. K., Papadopoulou, C. I., Vrochidou, E., Kaburlasos, V. G., Papadopoulou, M. T., Holeva, V., Nikopouplou, V.-A., & Dalivigkas, N. (2021). Social robots in special education: A systematic review. *Electronics, 10*(12), 1398. https://doi.org/10.3390/electronics10121398
5. Mustafa, M. Y., Tlili, A., Lampropoulos, G., Huang, R., Jandric, P., Zhao, J., Salha, S., Xu, L., Panda, S. P., Kinshuk, Lopez-Pernaz, S., & Saqr, M. (2024). A systematic review of literature reviews on artificial intelligence in education (AIED): A roadmap to a future research agenda. *Smart Learning Environments, 11*, 59. https://doi.org/10.1186/s40561-024-00350-5
6. Holmes, W., Bialik, M., & Fadel, C. (2019). Artificial intelligence in education: Promises and implications for teaching and learning. Center for Curriculum Redesign.
7. Paladines, J., & Ramirez, J. (2020). A systematic literature review of intelligent tutoring systems with dialogue in natural language. *IEEE Access, 8*, 164246–164267. https://doi.org/10.1109/ACCESS.2020.3021383
8. Ma, W., Adesope, O. O., Nesbit, J. C., & Liu, Q. (2014). Intelligent tutoring systems and learning outcomes: A meta-analysis. *Journal of Educational Psychology, 106*(4), 901–918. https://doi.org/10.1037/a0037123
9. Kulik, J. A., & Fletcher, J. D. (2016). Effectiveness of intelligent tutoring systems: A meta-analytic review. *Review of Educational Research, 86*(1), 42–78. https://doi.org/10.3102/0034654315581420
10. Alcorn, A. M., Ainger, E., Charisi, V., Mantinioti, S., Petrovic, S., Schandenberg, B. R., Tavassoli, T., & Pellicano, E. (2019). Educators' views on using humanoid robots with autistic

learners in special education settings in England. *Frontiers in Robotics and AI, 6*, 107. https://doi.org/10.3389/frobt.2019.00107

11. Belpaeme, T., Kennedy, J., Ramachandran, A., Scassellati, B., & Tanaka, F. (2018). Social robots for education: A review. *Science Robotics, 3*(21), eaat5954. https://doi.org/10.1126/scirobotics.aat5954
12. Ouyang, F., & Xu, W. (2024). The effects of educational robotics in STEM education: A multilevel meta-analysis. *International Journal of STEM Education, 11*, 7. https://doi.org/10.1186/s40594-024-00469-4
13. UNESCO. (2023). Guidance for generative AI in education and research. *UNESCO*. https://unesdoc.unesco.org/search/6e564eeb-7ead-4f94-9aff-a635e63d98a1
14. UNESCO. (2021). AI and education: Guidance for policymakers. *UNESCO*. https://unesdoc.unesco.org/ark:/48223/pf0000376709
15. Dong, L., Tang, X., & Wang, X. (2025). Examining the effect of artificial intelligence in relation to students' academic achievement: A meta-analysis. *Computers and Education: Artificial Intelligence, 8*, Article 100400. https://doi.org/10.1016/j.caeai.2025.100400
16. OECD. (2021). AI and the future of skills. In *Volume 1: Capabilities and assessments, educational research and innovation*. OECD Publishing. https://doi.org/10.1787/5ee71f34-en
17. Wang, S., Christensen, C., Cui, W., Tong, R., Yarnall, L., Shear, L., & Feng, M. (2023). When adaptive learning is effective learning: Comparison of an adaptive learning system to teacher-led instruction. *Interactive Learning Environments, 31*(2), 793–803. https://doi.org/10.1080/10494820.2020.1808794
18. Education Endowment Foundation. (2023). Evaluation report: Century Tech in UK schools. https://educationendowmentfoundation.org.uk
19. Kumazaki, H., Muramatsu, T., Yoshikawa, Y., Matsumoto, Y., Ishiguro, H., Kikuchi, M., Sumiyoshi, T., & Mimura, M. (2020). Optimal robot for intervention for individuals with autism spectrum disorders. *Psychiatry and Clinical Neurosciences, 74*(11), 581–586. https://doi.org/10.1111/pcn.13132

Future Innovations and Evaluation Methods in Robotics Education

1 Introduction

The increasing convergence of robotics with cutting-edge technologies such as artificial intelligence (AI), virtual reality (VR), augmented reality (AR) and the Internet of Things (IoT) is redefining the scope and direction of robotics education. In the wake of the fourth and fifth industrial revolutions, robotics is no longer confined to engineering labs but is evolving into a foundational discipline that integrates design thinking, computational intelligence and socio-technical awareness. This chapter takes a critical look at the emerging innovations and assessment frameworks that are reshaping the way robotics is taught, learnt and assessed in today's education systems.

Drawing on three decades of academic experience in robotics research and pedagogy, the chapter provides an in-depth analysis of technology-enhanced teaching strategies and their pedagogical implications. It explores how immersive, data-driven and interdisciplinary learning environments contribute to the holistic development of students' technical, cognitive and collaborative skills. It also explores how robotics education can dynamically adapt to global trends, changes in industry and socio-cultural contexts while ensuring inclusivity and lifelong learning.

By outlining the opportunities and challenges associated with the integration of cutting-edge technologies into robotics education, this chapter serves as a comprehensive guide for educators, curriculum designers, policy makers and researchers committed to future-ready learners and resilient educational ecosystems.

N. M. Thamrin et al., *Robotics in Education*,
SpringerBriefs in Applied Sciences and Technology,
https://doi.org/10.1007/978-981-95-3548-4_6

2 Emerging Technologies in Robotics Education

The landscape of robotics education is being fundamentally reshaped by new technologies that are redefining the way learning is designed, delivered and experienced. These innovations are not only expanding accessibility but also enriching the pedagogical techniques used in teaching robotics by combining theory and hands-on, experiential learning in new and meaningful ways. The integration of Artificial Intelligence (AI), Virtual Reality (VR), Augmented Reality (AR) and the Internet of Things (IoT) offer educators unprecedented opportunities to increase student engagement, deepen conceptual understanding and foster skills that meet the demands of the fourth industrial revolution. An exploration of these technologies reveals their unique contributions and the innovative strategies they enable in robotics education.

As summarised in Table 1, these technologies offer innovative learning pathways from personalised and adaptive systems to immersive and interactive environments that enhance both understanding and accessibility. Each technology contributes uniquely to overcoming spatial, cognitive or infrastructural limitations while supporting different pedagogical approaches such as gamification, project-based learning and kinaesthetic engagement. However, their implementation also comes with challenges, such as technological bias, cognitive overload and infrastructure requirements. A closer examination of each technology reveals its specific role, benefits and limitations in shaping the future of robotics education.

Table 1 Summary of emerging technologies in robotic education

Characteristic/ Technology	Artificial Intelligence (AI)	Virtual Reality (VR)	Augmented Reality (AR)	Internet of Things (IoT)
Learning pathway	Personalised, adaptive learning	Experiential understanding of core concepts	Improves understanding of complex processes	Remote monitoring and control of experiments
Accessibility	Sophisticated simulation environments	Overcomes geographical and infrastructure boundaries	Affordable and scalable solution	Overcome spatial restrictions and improve access
Pedagogical approach	Improves student preparations for autonomous system	Gamified learning experiences	Supports kinaesthetic and spatial learning	Fosters collaborative, distributed project-based learning
Challenges	Inclusion and bias in learning models	Manages cognitive load by simplifying operations	Cognitive overload and technological resistance	Scalability and infrastructure requirements

2.1 Artificial Intelligence (AI)

Artificial intelligence plays a central role in transforming robotics education by enabling personalised and adaptive learning pathways tailored to learners' individual profiles. Intelligent Tutoring Systems (ITS), for example, use algorithms that continuously assess student understanding and adapt teaching content accordingly to cater for different levels of ability and learning styles. This personalisation not only increases learning efficiency but also promotes a learner-centric environment where students receive timely and targeted support, significantly improving engagement and knowledge retention [1].

Beyond personalisation, AI enables the creation of sophisticated simulation environments in which learners interact with virtual robotic systems that embody the complexity of the real world, such as sensor integration, autonomous navigation and decision-making algorithms. Through these AI-driven simulations, students can gain hands-on experience with robotics operations and AI concepts without the need for expensive physical hardware, offsetting resource constraints, especially in underserved educational institutions. These virtual environments promote computational thinking by familiarising learners with machine learning architectures, pattern recognition algorithms and adaptive control strategies of the cognitive cornerstones of modern robotics [2].

Integrating AI into robotics curricula also improves student preparation for the development and deployment of autonomous systems. Since AI forms the intelligence that underlies robotic autonomy, curricula that combine AI and robotics provide learners with a comprehensive understanding of how intelligent agents recognise, plan and act in dynamic environments. This interdisciplinary approach encourages students to synthesise knowledge from mechanical design, programming and algorithm development and equips them with skills that are essential for innovation in automated systems [3].

However, the use of AI in robotics education also brings challenges, particularly in terms of inclusion and the risk of embedding biases in adaptive learning models. Careful attention to equitable design and transparent methods is required to ensure that AI-powered educational tools effectively help diverse groups of learners and support the democratisation of robotics education while mitigating widening inequalities [1].

2.2 Virtual Reality (VR)

Virtual reality (VR) is an innovative pedagogical tool for robotics education as it immerses learners in interactive three-dimensional environments that enable an experiential understanding of core robotics concepts such as kinematics, dynamics and control systems. The ability of VR to visualise otherwise abstract concepts in spatial and temporal contexts promotes cognitive assimilation far beyond traditional didactic methods [4].

The benefits of VR have been particularly evident during disruptions such as the COVID-19 pandemic, where access to physical lab spaces has been limited or impossible. VR-enabled remote labs offered a viable alternative, allowing students worldwide to participate in experimental robotics exercises with realistic simulations of hardware and programming tasks. This technology not only provided continuity in education, but also democratic access by overcoming geographical and infrastructural boundaries [5].

Furthermore, VR facilitates gamified learning experiences by structuring curricula into step-by-step challenges that mirror level-based game design. This approach motivates learners to persistently engage as they progressively build complex robotics skills and confidence in a supportive manner. The gamification inherent in VR encourages learner autonomy and problem-solving skills, contributing positively to motivation and knowledge retention, especially in challenging technical domains [6].

Importantly, VR interventions also help to manage cognitive load by simplifying and visualising complex robotic operations, reducing learner frustration and disinterest. Case studies where VR labs have been used in blended learning show positive student responses, improving both performance outcomes and the quality of the experience. The flexibility to repeat experiments from different angles, including angles not possible in physical labs, enriches the learning environment and supports reflective skills [7].

2.3 Augmented Reality (AR)

Augmented reality improves robotics training by superimposing virtual elements such as internal robot states, sensor readings and planned trajectories over real objects or environments. This overlay improves learners' ability to understand and interpret complex robotic processes that are normally opaque or difficult to visualise, such as path planning and sensor fusion, thus increasing explainability and learning efficiency [8].

The affordability and scalability of AR make it an attractive solution for educational institutions with limited resources. By utilising widely available devices such as smartphones and tablets, AR applications enable accessible interactive simulations that replace or complement expensive physical robotics kits. This democratisation of access supports inclusive education by enabling the teaching of robotics in different school contexts, including those with limited budgets or infrastructure [9].

From a pedagogical perspective, AR supports kinaesthetic and spatial learning through interactive experiences that engage multiple senses. Empirical studies have shown that AR can increase student motivation, concentration and knowledge retention by promoting immersive and tactile engagement with robotics content. In addition, AR promotes collaborative learning by enabling shared visualisations and interactions in physical or virtual spaces [10].

However, challenges such as cognitive overload, resulting from excessive presentation of information and technological resistance from teachers and learners need to be carefully managed to avoid compromising the learning benefits. Connectivity issues and complex technical requirements are further practical barriers to widespread adoption of AR in e-learning environments. Nevertheless, when integrated thoughtfully, AR is a powerful complement to traditional teaching methods by providing novel, effective learning experiences [10].

2.4 Internet of Things (IoT)

The Internet of Things (IoT) is fundamentally expanding robotics education by connecting robotic systems, sensors and learning platforms, enabling remote monitoring and control of robotics experiments in real-time. IoT-enabled remote labs allow students to design, debug and operate robots from geographically dispersed locations. This overcomes long-standing spatial restrictions and improves equal access to practical robotics training [5].

IoT infrastructures collect extensive multimodal data about student interactions and the status of robotic systems during exercises. Such data streams provide educators with detailed insight into learner behaviour and performance, enabling accurate and data-driven assessments that can inform adaptive teaching interventions [9].

In addition, the Internet of Things (IoT) fosters collaborative, distributed project-based learning by enabling synchronous communication and data sharing between team members working on robotics tasks, often across academic institutions or global teams. This connectivity mirrors practise in industry, where multidisciplinary teams coordinate via digital platforms to prepare learners for real-world workflows [11].

Scalability and infrastructural requirements are an ongoing challenge for the widespread adoption of IoT-enabled robotics education. Ensuring reliable network connectivity, managing data security and providing sufficient technical support are critical factors that need to be addressed to obtain the benefits of IoT integration [9].

3 Assessment Strategies for Robotics Skills

The assessment of robotics education requires comprehensive strategies that evaluate a full range of competencies, including technical skills, cognitive development and collaborative skills. By using versatile assessment methods, educators can holistically assess learners' progress and readiness for complex challenges in technical domains.

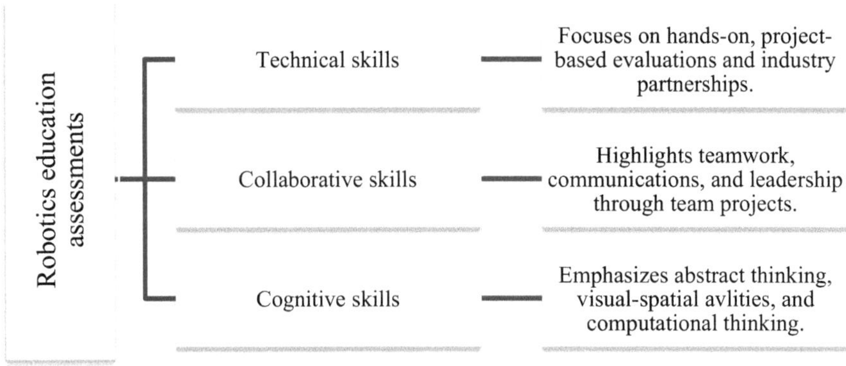

Fig. 1 Classification of robotic education assessments

In more specific terms, the assessment of robotics education involves a multifaceted approach that assesses students in three core areas: technical, co-operative and cognitive skills, as shown in Fig. 1. This comprehensive framework ensures that learners are not only proficient in practical, project-based tasks using industry-appropriate assessments, but are also able to work effectively in teams and demonstrate their communication, leadership and interpersonal skills. In addition, these assessments emphasise higher-order thinking by encouraging abstract thinking, visual and spatial reasoning and computational thinking skills that are essential for mastering the complexities of robotics and automation. In the following sections, the individual areas of assessment are explored in more detail, highlighting their importance and implementation in an educational context.

3.1 Methods to Evaluate Technical Skills

Technical competence in robotics is best measured through authentic, project-based assessments in which students are tasked with designing, implementing and refining robotic systems in a practical context. These hands-on assessments evaluate programming skills, hardware integration, system troubleshooting and creative problem-solving skills in a variety of scenarios [12].

Collaboration with industry partners plays an important role in aligning the technical assessments with real-world challenges. Such partnerships provide students with mentorship and feedback from professionals who are familiar with current technologies and workplace expectations. This improves the alignment of educational outcomes with labour market demands and facilitates iterative skills development [13].

The advent of remote and virtual labs expands assessment modalities by allowing instructors to monitor coding accuracy, interaction with simulations, and student diagnostic acumen in virtual environments. These platforms provide scalable, flexible assessment environments that can adapt to varying schedules and resource constraints [5].

Formative assessments embedded in practical modules allow for continuous tracking of student progress, enabling instructors to provide timely feedback and support iterative improvement. The structured use of assessment rubrics and quantitative metrics supports the objective measurement of technical competencies and effectiveness in problem-solving [12].

3.2 Methods to Evaluate Cognitive Skills

When assessing cognitive abilities in robotics, both quantitative and qualitative tools are used to measure abstract thinking, visual-spatial abilities and computational thinking. Technologies such as eye-tracking, in conjunction with traditional cognitive assessments, provide innovative methods to quantify learners' attention allocation, working memory capacity, and processing efficiency during robotic tasks [14].

Computational thinking remains central to robotics education and includes competencies in problem decomposition, algorithmic thinking, abstraction and pattern recognition. Assessment systems use standardised tests, classroom observations and teacher evaluations to provide comprehensive profiles of students' cognitive development in these areas [15].

Qualitative techniques such as thematic analysis of interviews complement the quantitative data and provide insight into how learners perceive robotics education, their motivation and the cognitive processes involved in learning. These combined approaches enrich the understanding of the nuances in students' cognition and the impact of instruction [14].

Longitudinal studies tracking cognitive growth indicate significant improvements in visuospatial working memory and abstract reasoning in learners participating in robotics programmes. This underpins the value of robotics instruction in promoting higher-order cognitive skills over time, from early childhood to higher education contexts [14].

3.3 Methods to Measure Collaborative Skills

Collaborative skills such as teamwork, communication and leadership are cultivated and assessed in the robotics classroom through team projects, competitions and co-operative problem-solving activities. Robotics competitions serve as authentic platforms for observing and evaluating collective dynamics and foster skills essential to a professional engineering environment [16].

The systematic use of detailed rubrics facilitates the structured assessment of creativity, leadership, collaborative problem-solving and interpersonal communication in a team context. Such tools enable consistent, transparent assessment of soft skills that are critical to both academic success and workplace performance [17].

Data analytics integrated into multi-user virtual environments provide real-time insights into interaction patterns, allowing instructors to monitor and measure the effectiveness of communication and quality of collaboration during robotics activities. These analyses support timely coaching interventions and targeted feedback [6].

Innovative instructional designs that combine flipped classrooms with collaborative coaching promote autonomy and teamwork while generating assessment data on group dynamics and individual contributions, enriching the assessment process [17].

Cross-cultural collaboration in robotics education broadens the perspective on teamwork and enhances the assessment of collaboration by incorporating different communication styles and problem-solving approaches, preparing students for globalised professional contexts.

4 Data-Driven Assessment and Feedback

Data-driven assessment methods in robotics education are becoming increasingly important for personalised learning and long-term academic success. Three prominent approaches, adaptive learning systems, learning dashboards, and longitudinal tracking, offer different functionalities, data inputs, and benefits that together improve educational outcomes as summarised in Table 2. Adaptive systems personalise content based on learner input, dashboards offer real-time progress monitoring and longitudinal tracking provides insight into retention of skills over time. Each method uses different data sources, such as sensor input, engagement metrics or developmental records, to optimise cognitive load, increase motivation or inform educational policy. However, for effective implementation, challenges such as privacy, the need for accurate misconception detection and long-term data collection requirements must be considered. The individual methods are explained in more detail in the following sections.

4.1 Adaptive Learning Systems and Analytics

AI-powered adaptive learning systems (ALS) epitomise the cutting-edge of data-driven educational assessment by dynamically adapting content delivery to a learner's ability, learning style and progress, improving the responsiveness and effectiveness of instruction [6]. By continuously analysing performance variables, ALS frameworks adjust tasks and feedback, optimise cognitive load and promote mastery.

Table 2 Comparison of data-driven assessment methods

Characteristic/ Approach	Adaptive learning systems	Learning dashboards	Longitudinal tracking
Functionality	Dynamically adapts content to the learner	Provides real-time insights into progress	Tracks skill retention over time
Data sources	Sensor inputs, user logs, assessments	Progress, performance, engagement data	Skill retention, cognitive development data
Benefits	Optimises cognitive load, promotes mastery	Encourages self-regulation, motivates effort	Provides insights for policy and curriculum
Challenges	Early detection of misconceptions is needed	Privacy, accuracy, and educator readiness are needed	Requires long-term data collection efforts

Multimodal machine learning further enriches these systems by integrating data from multiple sources, such as sensor inputs, user interaction logs, and assessment results, to infer learner understanding and engagement with greater precision. This technological sophistication enables early detection of misconceptions and learning deficiencies, allowing for proactive, targeted educational interventions [6].

This continuous cycle of data collection and analysis supports a formative culture of assessment in robotics education, allowing for iterative refinement of curriculum and targeted adjustments to instruction to meet the evolving needs of learners. Educators use data-driven insights to adapt support to maximise educational outcomes and learner motivation [18].

4.2 *Learning Dashboards and Visualisation Tools*

Real-time analytics dashboards in robotics education provide learners and teachers with actionable insights into progress, performance and engagement. Visualising learning progressions encourages student self-regulation by making abstract progress tangible and motivating further effort [1].

The integration of digital twins in VR and AR interfaces enables innovative visualisation of learning data in immersive and intuitive formats that improve understanding and enable more meaningful interaction with educational content. These interactive visualisations support differentiated instruction and facilitate inclusive educational practices by adapting the presentation to individual needs [4].

Despite their transformative potential, these tools pose challenges in terms of privacy, accuracy of interpretation, and educator readiness to use analytics effectively. Addressing these issues is critical to ensure the practical utility and ethical use of learning analytics in robotics education [5].

4.3 *Longitudinal Tracking and Impact Measurement*

Longitudinal assessment methods capture the lasting impact of robotics instruction on learner competencies by tracking skill retention and cognitive development over multiple semesters or school years. Such approaches provide evidence that learning robotics is associated with improvement in computational thinking, problem-solving skills, and collaboration skills over time [14].

Correlating longitudinal data with academic achievement and occupational outcomes provides valuable insights for policymakers and educators who can make decisions about curricula and identify pathways from education to employment. Mixed methods evaluations that combine quantitative performance measures with qualitative learner reflections provide comprehensive impact analyses [18].

These evaluations contribute to evidence-based policy making and curriculum reforms aimed at optimising the long-term benefits of robotics education and its alignment with emerging industry needs [11].

5 Long-Term Impact of Robotics Education on Career Paths

Robotics education has far-reaching implications across multiple dimensions, including advancing technology-driven industries, improving cognitive and social skills, and promoting inclusion and equity, as highlighted in Table 3. Through targeted skills development, it promotes advanced technological literacy, decision-making and teamwork, and provides access to technology-related career paths. The interdisciplinary curriculum integrates artificial intelligence and robotics while supporting academic perseverance and inclusive teaching practices. In addition, robotics education bridges the gap between academia and industry by teaching real-world, modern technologies, fostering emotional intelligence, and supporting policy frameworks that emphasise diversity and equal access. The following sections discuss each of these impact areas in more detail.

5.1 *Preparing Students for Tech-Driven Industries*

Robotics education is instrumental in fostering critical thinking, creativity and complex problem-solving skills aligned to the demands of Industry 4.0, encompassing digital manufacturing, automation and AI-enabled production environments. This focus ensures that graduates have the competences required for advanced technological tasks [12].

Table 3 Impact of robotic education

Characteristic/Impact	Tech-driven industries	Cognitive and soft skills	Inclusion and equity
Skill development	Advanced technological competence	Decision-making, teamwork, leadership	Access to technology-related career paths
Curriculum	Interdisciplinary integration of AI, robotics	Promotes academic persistence, lifelong learning	Accessible technologies, culturally responsive teaching
Industry exposure	Real-world contexts, modern technologies	Social and emotional competencies	Policies that emphasise equity and inclusion

Interdisciplinary curricula that integrate AI, robotics and software engineering equip students with the versatile skills they need to navigate the multi-layered technological landscape and drive innovation. This comprehensive education prepares students not only for technical tasks, but also for the challenges of collaborative development and systems integration [3].

By working with industry throughout their education, students are immersed in real-world contexts and enhance their employability through exposure to modern technologies and workplace cultures. Industry mentors and projects bridge the gap between academic preparation and professional expectations, facilitating a smoother transition into professional life [13].

5.2 *Development of Higher-Order Cognitive and Soft Skills*

Participation in robotics projects particularly improves executive functions such as decision-making, adaptability and strategic thinking, which are crucial in professional and academic environments. At the same time, soft skills such as teamwork, leadership and communication are enhanced, reflecting the multidimensional competences required by employers [18].

Meta-analytic evidence confirms the positive impact of robotics education on attitudes towards STEM, motivation and cognitive development. These outcomes promote academic persistence and lifelong learning, equipping students for rapid technological change [16].

Developing these higher-order skills is important not only for technical mastery but also for the social and emotional competencies required in dynamic work environments.

5.3 *Inclusion and Equity in Robotics Career Pathways*

Robotics education is an important way to facilitate access to technology-related career paths for underrepresented groups. Inclusive pedagogical design that incorporates virtual platforms, adaptive learning, and mentorship helps to equalise systemic inequalities in gender, geography, and socioeconomic status [19].

Efforts to bridge resource conflicts through accessible technologies and culturally responsive teaching strategies help to democratise robotics education and broaden participation. These initiatives promote social mobility and diversity in the STEM workforce, which are essential for innovation [20].

Policies that emphasise equity and inclusion in robotics education are critical to ensure that the benefits of technological advancement are shared across all populations and support equitable workforce development [1].

6 Continuous Improvement in Robotics Curricula

Improving robotics education requires a dynamic and inclusive approach centred on continuous curriculum improvement. This process is driven by three interconnected pillars: University–industry partnerships, curriculum alignment and professional development. Collaboration with industry gives students real-world experience and access to cutting-edge resources, while curriculum customisation ensures that course content remains relevant in the midst of rapidly evolving technologies. At the same time, ongoing professional development gives faculty the skills and confidence to use new tools and methods effectively. Together, these elements form a robust ecosystem that advances robotics education and prepares learners for future technological landscapes. In the following sections, each of these components is discussed in more detail (Fig. 2).

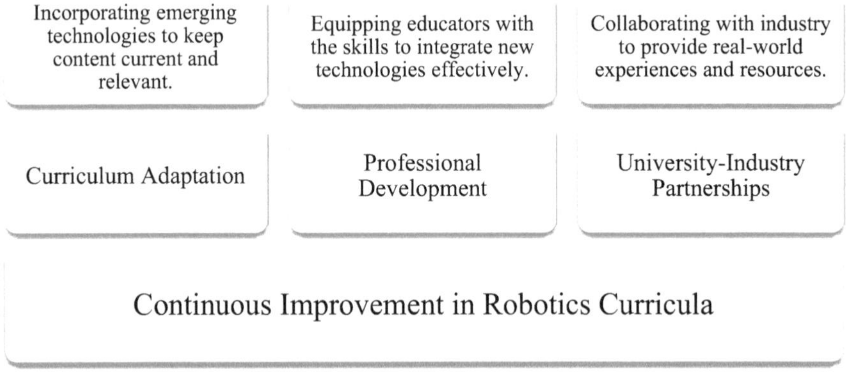

Fig. 2 Enhancing robotic education

6.1 Curriculum Adaptation to Emerging Technologies

Robotics curricula need to be continuously revised to incorporate advances in AI, VR, AR and IoT and ensure that content remains up-to-date and pedagogically sound. Modular, flexible course designs facilitate rapid updates and support the incorporation of new tools, methods and ethical considerations that reflect technological trends [21].

Curricula emphasise experiential, project-based learning that teaches both technical skills and critical thinking. The inclusion of interdisciplinary collaboration further enriches the learning experience and prepares students for complex challenges in the real world [12].

Managing cognitive load and maintaining learner engagement through technology-enhanced pedagogy improves instructional effectiveness and promotes deeper understanding and retention of content [22].

6.2 Professional Development for Educators

Sustained professional development is essential to equip educators with skills that enable them to effectively integrate new technologies into the robotics classroom. Training programmes develop digital pedagogical knowledge tailored to robotics and empower teachers to use and navigate innovative tools [22].

Support networks, mentorship models and communities of practise enable educators to share best practices and overcome resistance to technology adoption. These resources are critical to building trust and instructional flexibility among teachers [19].

Evaluations show that robust professional development has a positive impact on teaching quality and student learning outcomes, emphasising its importance in educational innovation strategies [20].

6.3 University–Industry Partnerships and International Collaboration

Strategic partnerships between academia and industry enhance the relevance of the curriculum by providing access to authentic projects, expertise and cutting-edge resources. These collaborations enrich curriculum content and introduce students to real-world problem-solving [6].

Cross-institutional and international collaborations broaden perspectives, facilitate knowledge exchange and harmonise assessment methods. Such alliances prepare learners for a global technological environment and support innovation in pedagogy and technology integration [11].

Sustainable partnerships help improve the employability of graduates and ensure that educational programmes evolve in line with industry trends [3].

7 Global Perspectives on Robotics Education

Global robotics education is characterised by a dynamic interplay of international trends, ongoing challenges and new best practices as highlighted in Fig. 3. On the one hand, the field is being driven by technological innovations such as artificial intelligence (AI), virtual and augmented reality (VR/AR) and the Internet of Things (IoT), which are reshaping the way robotics is taught and learnt. On the other hand, there are significant barriers, such as inequality, cultural biases and infrastructural limitations—that hinder equitable access and implementation in different regions. Successful models of robotics education, therefore, rely on inclusive pedagogy, blended learning strategies and comprehensive assessment systems to ensure inclusivity and effectiveness. In the following sections, each of these areas will be explored in detail to provide a holistic overview of the current global landscape of robotics education.

Fig. 3 Trends, challenges, and best practices of global robotic education

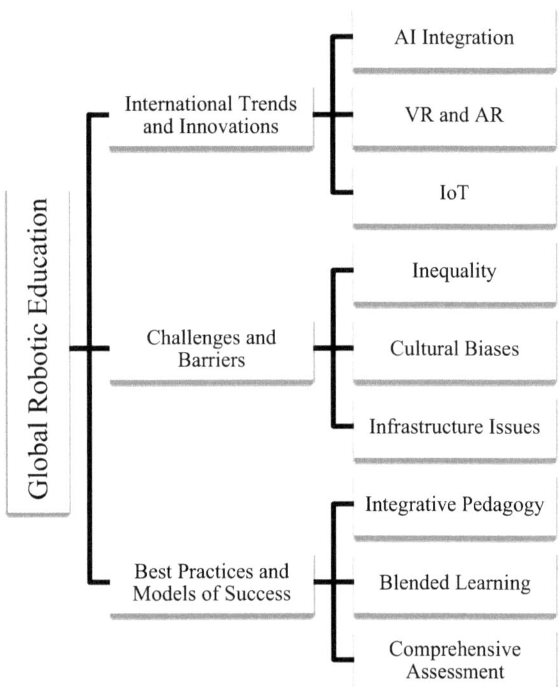

7.1 International Trends and Innovations

The global educational landscape in the field of robotics is characterised by the rapid integration of AI, VR, AR and IoT at all levels of education. This multi-technology embedding reflects a progressive shift towards interactive, immersive and technology-rich learning ecosystems that combine both hardware and software competences [11].

Efforts to increase inclusivity through virtual simulations and accessible platforms enable broader participation, mitigating traditional inequalities and promoting diverse learning groups [19].

In pedagogy, project-based learning and design thinking are increasingly used to foster creativity, collaboration and higher-order thinking skills that are essential for mastering robotics and technological innovation [16].

7.2 Challenges and Barriers to Implementation

Despite the encouraging developments, the problems of equality remain omnipresent. Unequal resources restrict access to robotics hardware and software, especially in low-income and remote areas, limiting educational opportunities [19].

Cultural biases influence educators' readiness and acceptance of robotics, necessitating contextualised professional development and culturally sensitive adaptations to the curriculum [19].

Infrastructure issues, such as limited connectivity, complicate the use of the technology. In addition, managing the cognitive load created by increasingly complex technological environments is critical to prevent learners from becoming overwhelmed and disengaged [4].

7.3 Best Practices and Models of Success

Best practices emphasise an integrative pedagogy based on constructivist learning theories such as Papert's constructivism and the maker movement, which emphasise creativity, autonomy and active learner engagement [20].

Blended learning approaches that combine physical robotics labs with virtual and remote access address the diverse needs of learners and alleviate resource constraints by promoting flexibility and continuity in instruction [7].

Comprehensive assessment strategies that incorporate technical, cognitive and collaborative dimensions ensure holistic assessment of learner development. Combined with sustainable professional development and global communities of educators, these practices ensure educational quality and innovation [6].

Collaboration between universities and industry supports the innovation of curricula and resources to adapt educational programmes to evolving workforce expectations and technological advances.

8 Conclusion

To summarise, the integration of AI, VR, AR and IoT into robotics education represents a profound development in both pedagogy and assessment, providing learners with essential skills for future technology-oriented professions. Robust assessment systems, including data-driven and collaborative evaluations, combined with continuous adjustments to the curriculum, underpin effective teaching and learner preparation. Addressing the ongoing challenges of equity, infrastructure and workforce readiness is critical to realising the full potential of robotics education worldwide. Strategic partnerships and global collaborations reinforce these efforts and promote innovation and inclusion. Taken together, these future innovations and assessment methods position robotics education as a cornerstone for developing the next generation of technologists, innovators and responsible leaders in an increasingly automated world.

References

1. Umak, B., et al. (2024). AI-based education tools for enabling inclusive education: Challenges and benefits. In *International convention on information and communication technology, electronics and microelectronics*. https://doi.org/10.1109/MIPRO60963.2024.10569714
2. Hijn-Neira, R., Pizarro, C., Borrs-Gen, O., & Cavero, S. (2024). AI-generated context for teaching robotics to improve computational thinking in early childhood education. *Education Sciences*. https://doi.org/10.3390/educsci14121401
3. Elizalde, L. B. M., Noguez, C. A., Cavazos, M. R. L., & Garza, L. R. S. (2025). Bridging AI, robotics, and software engineering: An interdisciplinary approach for learning emerging technologies. In *IEEE global engineering education conference*. https://doi.org/10.1109/EDUCON62633.2025.11016335
4. Tarng, W., et al. (2024). Application of virtual reality in developing the digital twin for an integrated robot learning system. Multidisciplinary Digital Publishing Institute. https://doi.org/10.3390/electronics13142848
5. Pang, D., Cui, S., & Yang, G. (2022). Remote laboratory as an educational tool in robotics experimental course. *International Journal of Emerging Technologies in Learning (iJET)*. https://doi.org/10.3991/ijet.v17i21.33791
6. Vassigh, S., Corrigan, S., Bogosian, B., & Peterson, E. Adaptive immersive learning environments for teaching industrial robotics. https://doi.org/10.54941/ahfe1004411
7. Wang, C., Belardinelli, A., Hasler, S., Stouraitis, T., Tanneberg, D., & Gienger, M. (2023). Explainable human-robot training and cooperation with augmented reality. https://doi.org/10.1145/3544549.3583889
8. Prokopiuk, N., & Falkowski, P. (2023). Applicability of augmented and virtual reality for education in robotics. https://doi.org/10.14313/jamris/3-2022/25

9. Alzahrani, N. M. (2020). Augmented reality: A systematic review of its benefits and challenges in e-learning contexts. Multidisciplinary Digital Publishing Institute. https://doi.org/10.3390/app10165660
10. Saputro, A. T., Fazil, A., Hidayat, N., & Novita, M. (2025). Evaluating the impact of TechnoVIT 2024 on AI, IoT, and robotics education for global challenges. *Cycle*. https://doi.org/10.63914/cycle.v2i1.36
11. Queiroz-Neto, J. P., Rodrigues, M. A. S., Pereira, M., & Gouvea, N. A. (2024). Integrating project-based learning and design thinking: An innovative approach to enhancing hard and soft skills in industrial robotics education. *Frontiers in Education Conference*. https://doi.org/10.1109/FIE61694.2024.10893483
12. Nayak, A. S., Hiremath, N. D., Patil, U. F.M, P. B., & Chickerur, S. (2024). Fostering higher cognitive skills through industry collaboration: A microcontroller-based approach. *Journal of Engineering Education Transformations*. https://doi.org/10.16920/jeet/2024/v37is2/24107
13. Liu, Y., et al. (2023). Effects of robotics education on young childrens cognitive development: A pilot study with eye-tracking. *Journal of Science Education and Technology*. https://doi.org/10.1007/s10956-023-10028-1
14. Kurniawan, W., Anwar, K., Jufrida, J., Kamid, K., & Riantoni, C. Personalized digital learning environment with differentiated instruction to foster computational thinking in robotics education. *Journal of Information Technology Education Innovations in Practice*. https://doi.org/10.28945/5427
15. Angelopoulos, P., Mitropoulou, D., & Papadimas, K. The contribution of open educational robotics competition to support STEM education and the development of computational thinking skills. https://doi.org/10.4018/978-1-7998-6717-3.ch022
16. Ortiz, E. M. L., et al. (2025). Enhancing engineering education through flipped learning and collaborative coaching: An empirical study on AI and robotics project development in higher education. https://doi.org/10.1109/EDUNINE62377.2025.10981414
17. Fan, O., & Xu, W. (2024). The effects of educational robotics in STEM education: A multilevel meta-analysis. *International Journal of STEM Education*. https://doi.org/10.1186/s40594-024-00469-4
18. Mistretta, S. (2022). Virtual robotics in hybrid teaching and learning. IntechOpen. https://doi.org/10.5772/intechopen.102038
19. Ali, N., Santos, I. M., AlHakmani, R., Khurma, O. A., Khine, M. S., & Kassem, U. (2023). Exploring technology acceptance: Teachers' perspectives on robotics in teaching and learning in the UAE. In *Contemporary educational technology*. https://doi.org/10.30935/cedtech/13646
20. Hamad, N. M. A., Adewusi, O. E., Unachukwu, C., Osawaru, B., & Chisom, O. N. (2024). Mentorship programs in robotics and their impact on student development. https://doi.org/10.26480/mjhrm.01.2024.69.77
21. Alimisis, D. (2021). Technologies for an inclusive robotics education. Open Research Europe. https://doi.org/10.12688/openreseurope.13321.2
22. Antonelli, D., et al. (2023). A virtual reality laboratory for blended learning education: Design, implementation and evaluation. Multidisciplinary Digital Publishing Institute. https://doi.org/10.3390/educsci13050528

MIX
Papier aus verantwortungsvollen Quellen
Paper from responsible sources
FSC® C105338

If you have any concerns about our products,
you can contact us on
ProductSafety@springernature.com

In case Publisher is established outside the EU,
the EU authorized representative is:
Springer Nature Customer Service Center GmbH
Europaplatz 3, 69115 Heidelberg, Germany

Printed by Libri Plureos GmbH
in Hamburg, Germany